SPRING FEVER

SPRING FEVER

*The Precarious Future
of Britain's Flora and Fauna*

PHILLIP GATES

HarperCollins*Publishers*

HarperCollins*Publishers*
77-85 Fulham Palace Road
Hammersmith, London W6 8JB

Published by HarperCollins*Publishers* 1992
9 8 7 6 5 4 3 2 1

Copyright © Phillip Gates 1992

The Author asserts the moral right to
be identified as the author of this work

A catalogue record for this book
is available from the British Library

ISBN 0 00 215892 2

Photoset in Linotron Baskerville by
Falcon Typographic Art Ltd, Edinburgh

Printed in Great Britain by
HarperCollinsManufacturing Glasgow

All rights reserved. No part of this publication may be reproduced,
stored in a retrieval system, or transmitted, in any form or by any
means, electronic, mechanical, photocopying, recording or otherwise,
without the prior permission of the publishers.

Acknowledgements

Throughout *Spring Fever* I have alluded to the published research of a large number of fellow scientists and naturalists; in many cases I have used results of their work to support my own interpretations of the likely effects of climate change. Without their original and painstaking research, I would not have been able to write this book. However, any errors of fact or interpretation are entirely my responsibility.

I am grateful to Sara Menguc of Murray Pollinger and Michael Fishwick of HarperCollins for advice and guidance, and to Juliet Van Oss for her editorial help and advice. I must also thank Professor David Bellamy, who generously agreed to read and comment on the manuscript. Most of all I must thank my wife, for her constant support and encouragement, and my children, whose curiosity about the natural world is a continuous source of inspiration.

PHIL GATES
Durham
October 1991

The author and publishers wish to thank the following copyright holders who have given permission for extracts to be used: Shiel Land Associates Ltd; International Thompson Publishing Services Ltd; Routledge; Penguin Books Ltd; Cambridge University Press; Michael Berkeley; Weidenfeld, George and Nicolson Ltd; A. P. Watt Ltd, on behalf of Michael Holroyd; Gaia Books Ltd; and the Random Century Group. Sources of chapter heading quotations: Chapter 1: David Bellamy and Sheila Mackie, *The Great Seasons* (Hodder and Stoughton, 1981). Chapter 2: Gilbert White, *The Journals of*

ACKNOWLEDGEMENTS

Gilbert White, ed. Walter Johnson (David and Charles Reprints, 1970). Chapter 3: John Stewart Collis, *The Worm Forgives the Plough* (Penguin, 1975). Chapter 4: Christopher Lloyd, *The Adventurous Gardener* (Allen Lane, 1983). Chapter 5: S. M. Walters and J. Raven, *Mountain Flowers* (Collins New Naturalist, series No. 33, 1956). Chapter 8: Oliver Rackham, *Trees and Woodland in the British Landscape* (J. M. Dent and Sons, 1976). Chapter 9: Thomas Hardy, *The Return of the Native*, Macmillan New Wessex Edition, gen. ed. P. N. Furbank (Macmillan, London). Chapter 10: E. B. Ford, *Butterflies*, 3rd edn (Collins New Naturalist, series No. 1, 1957).Chapter 11: Michael Berkeley, 'From Flint to Shale', *Second Nature*, ed. Richard Mabey (Jonathan Cape, 1984). Chapter 13: quoted in J. W. Wood, *The Illustrated Natural History*, (Routledge, Warne and Routledge, London, 1862). Chapter 15: Anthony Huxley, *Green Inheritance*, (Collins Harvill, 1984). Chapter 16: Richard Mabey and Tony Evans, *The Flowering of Britain* (Chatto and Windus, 1989). Chapter 18: James E. Lovelock, *Gaia, Journal of the Marine Biological Association* 69; 758 (Cambridge University Press, 1989).

Contents

1	The End of the Hosepipe	1
2	A Golden Age of Gardening?	14
3	A Plague on All Our Crops	27
4	A New Gardeners' Calendar	39
5	Sex, Alpines and Aliens	50
6	Frosts Are Slain	63
7	The Tangled Bank	74
8	An Ill Wind?	89
9	Out of the Flames	101
10	The Teeming Hordes	110
11	Nibbling Away at the Edges	125
12	The Parable of the Water Flea	137
13	Oceans of Ignorance	150
14	Fading Genes	165
15	The British Baked Bean	179
16	The Technological Fix	191
17	Wasted Acres	204
18	The Lie of the Land	219
	Index	237

ONE

The End of the Hosepipe

> The problem is that all too often natural history is regarded as something special that only happens way out in the country and its principles have nothing to do with our home. Pollution happens in other people's rivers, starvation is at least two worlds away and conservation only applies to the other guy . . . Get out and take a look, read the signs, acquaint yourself with the happenings that shaped the living landscape in which you find your roots.
>
> PROFESSOR DAVID BELLAMY, *The Great Seasons*

It was a minor act of juvenile delinquency that introduced me to the wonderful complexity of wildlife in freshwater ponds and streams. As restless ten-year-olds, brought up in the country, my friends and I occasionally lapsed into vandalism to enliven a rural existence where nothing very exciting happened for weeks on end. The tinkle of breaking glass sometimes provided the necessary thrills that were missing in small village life.

We often gravitated towards the mill pond, a large, dark pool overhung by willows, fed by crystal-clear water that ran off the South Downs and emptying into the brackish salt marsh at the top of a tidal creek. There was always a rowing dinghy tied to the sluice gate and eventually the temptation to set it adrift grew too strong.

The boat began to drift satisfyingly towards open water. The water mill had been converted into flats overlooking the pond and when we looked up we saw faces in the windows. We were in trouble. One of us would have to get the boat back.

I waded out to retrieve the craft and by the time I struggled back to the bank there was a reception committee waiting. But as we braced ourselves for a severe roasting we were in for a shock. It seemed that they had not seen us cast the boat adrift, but had only witnessed the aftermath and valiant rescue and were concerned for our safety. As a reward for our public spirited act, we persuaded them to let us use the boat; voyages on the mill pond quickly became a regular evening and weekend adventure.

It was a very long, kidney-shaped pond, shallow at one end but so deep at the other that it was only just possible to see the bottom on days when the water was perfectly still and clear. It smelled of dank mud and decaying water weed. Paddling around its edges, we discovered yellow flag iris, water forget-me-nots, brooklime, watercress, marsh marigold and flote grass. There were water boatmen and great diving beetles that hung from the surface, then crash-dived as you touched them with a grass stem. The voyages introduced me to the noise of dragonflies' wings, which rattled in flight.

By allowing the boat to drift silently it was easily possible to get closer to animals than I had ever managed before. We could get within a few feet of a water vole before it plopped into the water and disappeared, leaving a stream of bubbles and a swirl of muddy water. There were mallard and coot nests by the score and aggressive swans which had to be treated with respect. They could have sunk us.

The bottom was swathed in emerald-green weed and if you leaned over the stern of the boat in the shallows you could watch creeping caddis fly larvae moving over the mud in armoured suits of dead stems. Trout hung in the current near the inlet and on a good day we might even catch a glimpse of a marauding pike, cruising along the edge of the weeds.

It was a place of idyllic memories, of the kind that shape attitudes in later life. I moved away from the area many years ago and have no claims on it, but it is often a pleasure to recall the joys of discovery that I experienced on and around that millpond and its streams. So you may imagine my feelings when I returned during a recent period of drought and found

THE END OF THE HOSEPIPE

that it had become a bed of dried and cracked mud. It seemed like sacrilege.

Other similar old haunts, like the River Lavant where I had fished for minnows and dredged freshwater shrimps from the gravel, had suffered the same fate. What had been a sparkling river, full of feathery water crowfoot growing amongst chalk boulders and flints, had become little more than a sunken grassy path.

The disappearance of the millpond and the river was more than just a souring of my personal, treasured memories; it was a symptom of an insidious environmental problem. Up until then, like most people, I had considered summer droughts to be a minor price that had to be paid for the opportunity to build a reasonable suntan. It had not dawned on me that we were on the threshold of a period of change where lack of water would undermine the livelihood of farmers, leave gardens parched and withered and begin to change the whole face of the countryside.

Convincing the public that large areas of England are running out of water is proving to be as difficult as it would have been to persuade passengers on the *Titanic* that they were about to embark on a fatal voyage. It is almost beyond comprehension that most of the southern and eastern parts of our notoriously rain-swept islands will need to adjust permanently to the concept of living with water shortage. The usual reactions to this scenario are disbelief, quickly followed by indignation and outrage. But consider the facts.

In some parts of England, like Folkstone and Sheerness, recent rainfall has been lower than in Barcelona. The first winter drought order in history, banning hosepipes, car washes and watering of sports grounds, was imposed in the Bristol area by the Environment Minister in the middle of November 1990. Between March and October of that year the 1000 square miles around Bristol had experienced the lowest rainfall for a hundred and fifty years and the large Chew Valley reservoir remained woefully empty. As Christmas approached hosepipe bans were also still in place in Southern, Thames, and Anglian water regions. In 1990 twenty million people suffered restrictions on their use of water.

In the spring of 1991 the managing director of the Cambridge Water Company, which serves the fertile agricultural areas between north Norfolk and north Essex, described the current drought as 'the worst this century'. Hosepipe bans in the area were not due to be lifted until the spring of 1992 at the earliest and the company warned gardeners not to waste money on bedding plants or laying a lawn. No water would be available to keep either alive.

Cambridge Water prepared special leaflets advising gardeners of the best methods for dealing with droughts. This is something that is familiar to Californian householders, who can now pay to have their dead lawns dyed green during long, dry summers, but no one ever dreamed that it could happen here.

The summer droughts of 1989 and 1990 provided the biggest trauma for avid gardeners in living memory. Thousands spent a large part of their evenings and weekends struggling to keep wilting vegetables and flowers alive with buckets of bathwater. Those seen using a hose during daylight hours ran the risk of being reported to local water authorities by outraged neighbours. Even surreptitious watering activities after dark were fraught with danger: snooping helicopters patrolled the worst-hit areas, on the lookout for lawns that were suspiciously green, a telltale sign of illicit use of sprinklers. Serious gardening suddenly became a clandestine activity.

The length and extent of recent droughts provided a foretaste of the conditions that we might experience on an annual basis by the middle of the next century. Water will become an expensive commodity and water meters will encourage householders to use it frugally.

These economic and domestic consequences pale into insignificance compared with the effects of water shortage on the environment in general. Rainfall and temperature are the two interacting factors that exert the major controlling effect on the distribution of plants and animals; shortage of water is already having a disastrous effect on river and wetland ecosystems in the worst-affected parts of the country. Whole river systems are disappearing from the map.

All over the south and east water levels are rapidly falling and rivers are drying up. The root cause is not changing climate but over-exploitation of our water resources. An enormous population density in regions where rainfall is naturally low in summer has led to rapidly escalating demand for water. Rising standards of living are to blame: car washes, dishwashers, automatic washing machines and increased numbers of baths and toilets have all added to the demand. Increased leisure time has generated a gardening boom which needs massive volumes of water to sustain it: lawn sprinklers use as much water in an hour as one person would use in a week.

In the south and east, deep extraction from aquifers has lowered the water table and several years of low rainfall have not recharged groundwater supplies. Some springs which fed river headwaters have ceased to flow. As long as groundwater is pumped at its present rate, river systems and their associated wildlife will continue to decline.

The fate of at least forty rivers now hangs in the balance. Included in these are such famous waterways as the Rivers Hamble, Frome, Meon, Little Stour, Wharfe and Derwent. Many of the upper stretches of rivers that are fast disappearing were once numbered amongst England's finest and most famous trout streams. The fish, and all of the associated wildlife, are threatened.

Low winter rainfall and high summer temperatures have prevented recharging of aquifers and water companies are now firmly of the opinion that consumers will no longer be able to enjoy unfettered use of what has traditionally been one of our most abundant natural resources. The days of the hosepipe are numbered, even if future levels of rainfall rise substantially in the south and east. Water meters will soon be here to stay.

The events that brought this creeping water crisis into focus were the series of summer droughts that dominated the climate in the late 1980s. This, the hottest decade in recorded climate history, contained six of the seven warmest years ever measured in England. 1990 was the warmest year ever and had the driest spring since 1893.

The climate in the 1980s was more than just a hiccup in our weather systems. Climatologists agree that we are now experiencing a shift in weather patterns resulting from man-made changes in the atmosphere. Average temperatures are rising. The weather is changing on a long-term basis. There is no doubt about that. The problem now is to determine how fast the changes will happen and how great the increase in temperature will be.

Some conservative estimates suggest a rise of one degree centigrade by the year 2040, which would be the fastest average temperature rise in the last ten thousand years on a planetary scale. More pessimistic forecasts indicate that the temperature might rise by four degrees centigrade by the end of the next century, the fastest rise in forty million years.

Even in the light of the most optimistic of these scenarios, the future prospects for water supplies in the south and east, where the climate is expected to become warmer and drier, are grim. The possibility that climate amelioration will turn what is currently a serious inconvenience into a major environmental problem is fast becoming a probability.

The reason for our vulnerability to this kind of climate change has much to do with the poorly planned way in which natural resources have been exploited. We have used up underground reservoirs to such a degree that there is no longer enough water left to carry us through periods of low rainfall. A warmer, drier climate in southern and eastern England will exacerbate an existing problem, adding an additional burden to an already precarious situation. Tackling it will be, in effect, like trying to turn down the flame under a pan of water which is in danger of boiling dry and finding that the gas tap has stuck.

So we are exploiting some parts of our water resources to the maximum and the long-term solutions will be very expensive.

Water will need to be piped from the wet north to the dry south, where its use will be closely monitored.

The future prospects for freshwater supplies mirror the ways in which climate amelioration will affect our flora and fauna. We treat our wildlife resources in much the same way as we treat our water. They are taken for granted until a crisis occurs. We store plants and animals in scattered, vulnerable reservoirs, variously known as nature reserves and Sites of Special Scientific Interest. Elsewhere, as with our water systems, wildlife habitats are abused, polluted and destroyed. And as with water reserves, the status of our flora and fauna is already precarious, making them extremely vulnerable to climate change.

The future course of climate amelioration is impossible to predict with any accuracy. Computer models deal in averages and general weather circulation patterns. Ask climatologists for their predictions for Britain's future weather and they will say that the climate will probably be even less predictable than it is today. 'Maybe' is a word they use often.

Analyses of the situation published by the Meteorological Office say that we are moving towards a climate where we will experience between 5 and 15 percent more rain, most of which will fall in winter. This change in the seasonal pattern of rainfall has already been a feature of the 1980s.

Winters will be milder, with less snow and fewer frosts, while summers will become hotter and drier. These generalizations hide major regional variations. Much of the increased rainfall will probably occur in the north and north-west, while the Home Counties will assume a mediterranean-type climate somewhat like that of the south of France.

Perhaps the only outcome of climate change that has been predicted with any certainty is that average temperature will rise. How this will affect humidity, surface evaporation and cloud formation, precise distribution of rainfall or occurrence of gales is beyond the present predictive capabilities of meteorological science. Quite simply, there is neither enough information nor adequate computing capacity to forecast these changes with any degree of accuracy.

The Great Gale of 1987 is still fresh in the memories of many and has prompted questions as to whether such extreme events will become common in the future. Changes in the surface temperature over land and sea precipitate mass movements of air, so any disturbance of surface temperatures is almost certain to change the strength and direction of winds. Although no one can be certain, the gales that lashed the south of England at the end of the 1980s could be a portent of things to come as surface temperatures change.

One of the less contentious predictions is that sea levels will rise, through the influence of melting glaciers and polar ice and because of the tendency of water to expand as it gets warmer. Recent evidence suggests that melting of polar ice will be slower than anticipated; indeed, more snow may fall at the poles.

As with every other aspect of climate amelioration, the scale of sea level changes is controversial. Even some of the more cautious estimates, of a rise of 15 cm. in the next fifty years, would have profound consequences for Britain's coastal wildlife habitats, like cliffs, salt marshes, estuaries and sandy beaches. If we opt to stem the rising tide with sea defences, the wildlife that occupies the shores of Britain will have nowhere to retreat to. Coastal wildlife will have its back to the wall, or, more precisely, the sea wall.

Some alternative scenarios paint a more alarming picture of climate change. The oceans act as a climatic thermostat, rather like storage radiators in that they damp down daily, seasonal and annual fluctuations in temperature. About a quarter of the sun's heat output that reaches our shores is first absorbed in the surface waters of the North Atlantic, before being released to warm the air that moves over the land.

Atmospheric temperature has a major influence on the course of ocean currents. The North Atlantic Drift, a weak offshoot of the Gulf Stream, plays a key role in maintaining the mild, wet climate along the west coast of Britain and Ireland. What might happen if its course changed?

The likely outcome, even if its course were to deviate northwards only slightly, would be that Scotland and the Atlantic coast of Britain would experience colder winters; Scotland's climate might resemble that of Norway.

So there are numerous alternative and confusing prophecies for the course of climatic change. But, in general, there does seem to be a consensus that there will be a north-south divide with warmer, drier conditions in the south and east and milder, wetter conditions in the north.

For much of our wildlife, confined to reserves where it is currently in equilibrium with temperature and rainfall patterns, any rapid, major change is a potential disaster. The future of our existing flora and fauna depends on species being able either to migrate and track changing environmental conditions, or to evolve at a rate that will allow them to keep pace with change.

As we shall see, there is little evidence that many species will be able to do either with much success.

There are certain to be major changes in the flora and fauna in areas of Britain that are most severely affected by climate amelioration. Whether these will be for better or for worse depends on your point of view.

Those who view our current wildlife in the same way that they might view a treasured personal possession, like a car or a house, will perceive any unpredictable and uncontrolled change as an unmitigated disaster. But to those who take a strictly biological viewpoint, climate change might herald an extraordinarily interesting phase in the natural history of our islands, when we will gain more species than we will lose. Warmer summers and fewer frosts could give many plants and animals that are currently rare natives or introduced aliens a chance to expand their range. Our flora and fauna, which has been evolving continuously under the influence of climate fluctuations for millions of years, could actually become more varied.

The future of our native plants and animals during a phase of rapid climate change has to be viewed in the context of past weather changes. Over the last sixty million years the flora of the British Isles has undergone massive transformations, as cycles of warm weather have alternated with phases when glaciers crept over the landscape, freezing sensitive plants to death.

Analysis of the fossil plant deposits in London clay shows that 73 percent of the plant species that are represented in it have living relatives in Malaya; at the time when these geological deposits were laid down, what is now London was a landscape of tropical brackish estuaries, which looked startlingly similar to those in present day Vietnam.

This tropical flora was followed by a transition to one which was more characteristic of warm temperate climates, represented by trees like magnolia. The plant life during this period was rather like that of present-day Burma, Western China and Japan.

As the climate cooled still further these exotics were replaced by species which we now associate with North America – wingnut, tulip-tree and conifers like hemlock. The climate continued to deteriorate and as mass extinctions of frost-sensitive species took place the British flora took on a distinctly European look.

Then came successive cycles of glaciation and warm interglacials, bringing with them an ebb and flow of species; by the time the last glaciers retreated bare areas of landscape were undergoing colonization by species from parts of Europe that had remained unscathed by ice. The composition of the present-day flora was finally set when the land bridge to Europe flooded, around 5500 BC.

The end of the last Ice Age was followed by a series of climate changes, which can be inferred from patterns of fossil pollen deposition in peat bogs. Peat deposits are like time machines: take a core downwards from their surface and the zones of pollen embedded in the successively deeper layers tell the story of past plant distributions. Since these are known to be related to warmth and water, dissecting peat bogs is like replaying a video recording of past climate.

The pollen record reveals a distinct sequence of changes. As the ice melted, the sub-arctic conditions gave way to a climate that was warmer and drier then before. This was followed by the Atlantic period, which was warmer and wetter, and then a period of continental conditions that was warm and dry. For the last two and a half thousand years the climate has been deteriorating and has generally become colder and wetter.

Each of these climate zones lasted between 2000 and 2500 years, and the transition between the warm and wet and warm and dry phases, about 5000 years ago, is what palaeobotanists call the climatic optimum, corresponding to a period in which frost-sensitive species thrived over much of the country. This phase produced what is known as the climatic climax vegetation, when most of Britain, including the high Pennines, was warm and moist enough to support forest, the final stage in the natural progression of vegetation changes in a stable climate. Then, only the mountain tops were devoid of trees.

Since then the climate has deteriorated and the vegetation of the British Isles has come under human influence. Neolithic farmers began settled agriculture and Bronze and Iron Age farmers cleared forests. Up to a point, this deforestation must have been beneficial, since it created open habitats where a great diversity of species could thrive that would not tolerate the shade of tree canopies.

During recorded history there have been cyclical fluctuations in temperature patterns; phases which were warm enough to allow the widespread cultivation of sun-loving crops like cannabis and cold enough to freeze the Thames. During the Little Ice Age, between 1500 and 1850, glaciers in the Alps and the Arctic advanced again and produced sequences of exceptionally harsh winters, allowing ice fairs to be held on the Thames. The last of these took place during January and

February 1814 and since then winters have become markedly warmer.

During these comparatively recent changes in climate there would certainly have been variations in the abundance of plant and animal species, but the science of biology was too poorly developed for them to be systematically recorded and measured. So up until now our knowledge of the effects of changing climate on plant and animal life has depended on recent fragmentary and often anecdotal information and on the careful analysis of scattered subfossil remains. Now, as the climate enters a period of rapid, man-made change, we have a unique opportunity to witness and perhaps even understand the mechanisms that have shaped the flora and fauna of the British Isles.

In the last few years it has become apparent to professional biologists that a close study of plant and animal life offers us the best means of monitoring the effects of climate change. Even minor fluctuations in temperature have an impact on their patterns of growth and behaviour. A small change in temperature has the same effect on plants as an increase in interest rates on the economy of a nation. The effect is cumulative over a period of time and, although the change may be small, the net effect at the end of the accounting period can bring ruin or salvation. Rates of annual growth may change and competitive interactions be disrupted. Such changes have a destabilizing effect on systems that have settled into temporary equilibrium. So plants, animals and ecosystems are probably the best barometers of change.

A mild winter encourages early bud burst in some trees and early emergence for some hibernating animals. It also allows many plants and animals to survive winter more successfully or even continue to breed throughout the cooler months of the year, so that their numbers increase. Higher summer temperatures are a positive benefit to insects like butterflies and can encourage earlier and more prolific seed set in flowering plants. At the same time, dragonfly and frog populations may be damaged by earlier drying of ponds, while fragile plants like mosses will suffer from severe drought. The response of the flora and fauna will be the first indication of the longer-term effects of a modified climate.

We are fortunate in Britain in having a very long tradition of gifted natural historians. It is probably no exaggeration to claim that our flora and fauna has been studied for longer and in greater depth than that of any other country. A climate change may also trigger a renaissance for amateur naturalists, presenting them with opportunities to contribute to knowledge of the interrelationships between plants, animals and the climate.

Throughout this century biology, the last of the sciences where the amateur could contribute, has become relentlessly more professional as observational skills have been replaced by laboratory experimentation and elaborate instrumention. It has become, like the rest of science, a discipline which affects the everyday life of many but which is practised by an elite few. With the onset of a period of rapid climate amelioration, the skills of amateur naturalists may once more come into their own. Our observations may become an integral part of the record of the most significant environmental changes in modern history.

These are exciting, disturbing times to be alive for anyone who has an enquiring mind, a keen sense of observation and an interest in the natural world. We are all involuntary participants in a giant experiment with no clearly predictable outcome. By observing and interpreting the changes that occur, as they occur, we may gain a better insight into our own future and our own real position in the complex pattern of nature.

TWO

A Golden Age of Gardening?

> The frost has killed the tops of the wallnut shoots, & ashes; & the annuals where they touched the glass of the frames; also many kidney-beans. The tops of the hops, & potatoes were cut-off by this frost. The tops of laurels killed. The wall-nut trees promised for a vast crop, 'til the shoots were cut off by ye frost.
> GILBERT WHITE, *journal entry for 25th May 1776*

From the moment that I first saw *Callicarpa rubella*, I knew that I had to have it.

For much of the year this nondescript shrub from the foothills of Assam would not attract a second glance. In autumn it will stop you in your tracks. Then, the twiggy stems are festooned with dense clusters of lilac-purple berries, of a colour that is almost unique amongst decorative fruiting shrubs.

I should have known better but I was gripped by horticultural avarice. *Callicarpa* is definitely a plant for sheltered parts of southern England and expecting it to survive, let alone perform, in my cold corner in West Durham was a triumph of optimism over good sense. Despite being planted against a south-west facing fence, it failed to produce a single live shoot at the end of its first winter.

I suppose that every keen gardener has a fund of similar stories. Most of us, at one time or another, have bought plants that we know will struggle to survive our invigorating winters. Better to have loved and lost, though, than never to have loved at all.

But now, things may be about to change. The climatic shifts

that are talked about should be enough to change the face of British gardening, tilting the balance in favour of those who have a pioneering horticultural spirit. The demise of the hard winter frost could turn this into a land fit for subtropical and Mediterranean plants to live in. Let me whet your appetite.

Flannel bush is a sumptuous shrub with an unflattering name which comes from sun-drenched California; its more distinguished Latin epithet, *Fremontodendron californicum*, suits it much better. It is a wonderful shrub for a hot sunny wall, where its grey-green foliage can sunbathe against hot brick and its exotic, buttercup-yellow flowers can open against a terracotta-coloured background. Even the large, fleshy buds are spectacular. They take an age to develop and when they are swollen and poised to open it seems as though they should burst with an audible 'pop'.

Sadly, in our present climate, *Fremontodendron* is really only suited to the milder areas, and then only against a sunny wall. Mine is confined to a large pot and spends the winter secure in a greenhouse. What a pleasure it would be to have sufficient confidence in a mild climate to be able to train this plant over the outside of the house.

Today there are a number of select plants that only gardeners in the south and west dare cultivate, which should become viable for a wider cross section of the gardening public as the climate improves. Typical of these is *Convolvulus cneorum*, a Mediterranean shrub with delightful silky, silvery-grey foliage which produces a continuous display of glistening white, trumpet-shaped blooms throughout the summer. The textures of the leaves and flowers are so inviting that it is hard to resist touching it. Like the velvety leaves of *Stachys lanata*, which comes from a similar hot and dry habitat, and the polished mahogany perfection of the trunk of *Prunus serrula*, it is a plant to fondle.

Understandably, it is also a plant that many gardening journalists rave about, but one which relatively few people successfully grow, since it comes from sun-drenched limestone hillsides and can only survive here with the protection of a wall in the mildest of regions. As summers become warmer, and more importantly, as winters lose their frosty caress, gardeners in more northerly counties should be able to appreciate the delights of *Convolvulus cneorum*, which looks particularly attractive when it sprawls over paving slabs in the company of other Mediterranean species, like aromatic thymes and *Cistus* species from the Garigue.

Plants like *Fremontodendron californicum* and *Convolvulus cneorum* from mild climates will add a new dimension to British gardening by the middle of the next century. If you have ever hankered after growing frost-sensitive exotics like Australian bottlebrushes or even the nearly-hardy hebes and lobster claw (*Clianthus puniceus*) from New Zealand, by the middle of the next century you should be able to plant with confidence. Or more to the point, you will be able to leave these vulnerable species in the open ground throughout the winter. No more laborious digging up or taking cuttings of *Gazanias* or *Penstemons* or Jamaica primroses (*Agryranthemums*) and overwintering them in the greenhouse or on window-ledges. Even the dahlias and the pelargoniums should become hardy perennials by the tail end of the twenty-first century.

Mediterranean cistus species, half-hardy perennial salvias from South America, Australian eucalyptus or a fuchsia hedge are all temptations that will be worth succumbing to, even if you do not have a sunny, well-drained border backed by a wall. But do not think of planting any of these just yet.

The essential characteristic of rising global temperatures is that although averages will gradually increase, severe fluctuations will continue to occur above and below this figure. How severe the oscillations will be is an open question. Rare but savage late frosts can do untold damage. The effect is particularly devastating after a mild winter and early spring, when early growth can be severely scorched.

The springs of 1989, 1990 and 1991 provided a classic example, when late frosts devastated the plum crop in the Vale of Evesham. After three successive years of disaster the plum growers began to doubt the viability of the crop. The same frosty spells scorched the shoots on my roses and killed all the precocious flower buds on my *Spirea x arguta*.

In general, alternating temperature extremes tend to make plants less hardy. It is this element of uncertainty that will be most worrying for adventurous gardeners who try to exploit a changing climate by planting half-hardy species outside. Anyone toying with the idea of planting an Indian bean tree or a Chilean *Eucryphia*, both of which might eventually become a much commoner sight in a warmer Britain, should be prepared for setbacks. If the clematis show signs of early flowering, watching the weather forecasts and being ready to dash out and cover its flower buds will be more important than ever.

In many ways a mild winter will be a mixed blessing for gardeners. It may mean that winter frost damage is less of a problem, but at the same time it will speed up development of flower buds and young, succulent, tender shoots. The winters of 1989 and 1990 were barely worthy of the name in much of England and the balmy days in late February brought some plants into bud six or eight weeks earlier than normal. Many gardeners paid the penalty when frosts struck in April. Marginally hardy plants, like tree paeonies, suffered badly.

One characteristic of tender plants that compounds the problem of deciding when to risk planting is that they acclimatize to frost. Young walnut trees are killed by severe low temperatures, whereas mature trees only suffer damage to small branches. Likewise, the Chilean southern beech, *Nothofagus*, is far less hardy as a sapling than it is in old age. Nursing juveniles through the confusing transitional weather conditions created by climate amelioration will require skilled gardening.

One way in which horticulture might deal with these conflicts created by higher temperatures and less predictable seasons is to reselect new varieties of existing garden species. Many important horticultural species have a wide natural geographical range, and by searching out individuals from the margins of these areas there is a good chance that varieties can be produced with greater environmental tolerance. A large number of cultivated plants from exotic locations have been introduced into cultivation from small collections made from a few locations within their range. The original collections were almost certainly made on the basis of the appearance of flowers or foliage, with no regard whatsoever for useful characteristics like temperature tolerance.

There is great scope for making new collections from the wild and exploiting such natural genetic variation from the edge of species' ranges. Equally, plant breeders can help by selecting clones from their breeding programmes on the basis of their resistance to stresses. Once these more adaptable varieties are identified, micropropagation, a remarkable development in horticultural technology, will ensure that they reach our gardens with the minimum delay.

Micropropagation is a means of multiplying plants at a rate that would have been beyond our wildest dreams a decade or two ago. The technique involves isolating a small piece of tissue from a desirable plant – perhaps a shoot tip or a bud from the base of a leaf, sterilizing it and then culturing it on a medium that contains a cocktail of plant hormones.

The tissue forms multiple shoots and each of these can be isolated so that the process can be repeated *ad infinitum*. Under favourable conditions one shoot can become ten in a fortnight, a hundred in a month or a million within three months. Once enough shoots have been produced, they can be transferred to a medium containing hormones that trigger root formation and from there they can be weaned in a glasshouse.

This technology is already used to mass-produce many of the herbaceous and woody perennials, and if you have bought a *Hosta* or *Choisya* recently, the odds are that it started life in a sterile nutrient jelly. When the climate changes and new varieties are needed quickly, this will be the means by which they will be made available to the gardening public.

If enrichment of the atmosphere with carbon dioxide means that hot summers will become the norm and harsh winters will be banished, then British gardens will take on a far more exotic character in high summer by the end of the next century. But long before then, winter itself should also become a much more colourful time of the year.

For many gardeners, winter-flowering shrubs are something of an afterthought. We tend to regard spring and summer as the periods of the year for gardening, and in winter a period of dormancy sets in, for gardeners and their plants. This is a pity, because there are some exquisite winter flowering shrubs available that can lift the spirits in the mid-January gloom. I have a *Mahonia japonica* outside my window, whose long trusses of small yellow flowers release a wonderfully heady lily-of-the-valley scent from the dark depths of February until the arrival of spring. Visually, it is not a particularly exciting plant, but the memory of its fragrance is enough to entice me out of doors on the most miserable February day for a long and satisfying sniff.

One legacy of warmer winters should be that these off-season shrubs will flower for longer and bloom more prolifically. Many should perform earlier; so much so that we will find ourselves regularly supplementing the Christmas holly decorations with vases of witch hazel (*Hamamelis*), wintersweet (*Chimonanthus praecox*) and perhaps even a few sprays of *Rhododendron dauricum*.

For gardeners who would like to exploit the benefits of warmer winters, there are many unexplored botanical possibilities. The honeysuckle family, with over two hundred members, embraces several winter-flowering species which will fill the air with fragrance on a still day. Desirable winter-flowering species include the aptly-named *Lonicera fragrantissima*, *L. standishii*, their hybrid, *L. x purpusii* and the much less common *L. setifera*. The last two species flower on bare twigs, which gives them a particularly oriental delicacy. All four flower earlier and more prolifically in mild winters.

Unlike the exotic, summer-flowering species which will creep into our gardens slowly as the climate warms, these winter bloomers are already adapted to bad winters, so we need have few reservations about introducing them into our gardens at the earliest opportunity. Since they are able to withstand the worst that our present winters can throw at them, the vagaries of a changing climate, with late frosts, should not do them too much harm.

Nevertheless, there are winter flowerers which are surprisingly tender. The best of these is *Abeliophyllum distichum*, a slow-growing jasmine-like plant to train against a wall, preferably under a window where its fragrance can waft into the room. It is susceptible to severe winter weather and so needs to be sited with the possibility of late, vicious frosts firmly in mind. The evergreen climbing *Clematis cirrhosa* from Majorca is also a winter flowerer with large, yellowish-cream flowers which are spotted with brown. This too is fairly tender, so cannot be expected to perform to its best abilities in the more northerly areas of Britain until the climate has settled into a reliable pattern of higher average temperatures throughout the year. When it does, the sight of a flowering clematis rambling over a trellis in February will confirm beyond doubt that our climate has well and truly changed.

All of these winter flowers will need to be treated with some kindness and consideration, because their winter performance will also depend on summer growth. Most flower on wood that has been made in the previous summer, so maintaining a supply of water in the dry months will be essential for spectacular flowering in the darkest months of the year.

In the new climate there will be no gain without pain.

In a warmer Britain the seasons in the garden will become increasingly blurred and the range of plants that will survive as hardy perennials will steadily expand. But it is not just the year-round performance of ornamental shrubs that will be enhanced. Every garden plant, from the humblest herb to the cucumbers in the glasshouse, will be affected to some degree.

Provided that adequate levels of soil moisture could be

maintained, reliably hot summers should do wonders for many plants from hot, dry climates that are unreliable performers at present. Everlasting flowers, like *Statice* species, *Helichrysums*, *Xeranthemums*, *Rhodanthe* and *Ammobium* will revel in long, hot summer days. Low humidities will make it easier to harvest and dry them rapidly in the kind of airy conditions that encourage them to retain their vibrant colours.

Most aromatic herbs, like basil, marjoram, lavender, rosemary and thyme, thrive in higher temperatures, producing more of the volatile oils that give them their distinctive fragrance and flavour. The presence of these oils in the foliage may actually be directly related to the drought tolerance of these species, which probably explains why the oil levels rise as the temperature increases.

Other plants benefit from a baking for different reasons. Mediterranean bulbs, like species tulips and *Alliums*, flower better after a good roasting in summer heat. Some irises bloom more profusely if their surface rhizomes are drenched in sunshine.

The spring flowers of the eastern Mediterranean are one of the most breathtaking spectacles of the botanical world. While I was working on the breeding of lentil crops in the mid-1980s, I took the opportunity to explore the flora of a small rocky hill called Tel Hadya that rises from the flat agricultural land of central northern Syria near Aleppo. The rocks on these stony outcrops become almost too hot to touch in the summer, and little or no rain may fall between April and October, but as the last spring rains die away they blaze with colour.

The crevices between the rocks are filled with blood-red anemones, miniature wild *Allium* species, asphodels, grape hyacinths and annuals like wild calendulas. In the more fertile agricultural soils, wild tulips and gladioli grow as weeds and are left in piles around fields as farmers pull them out of their crops, as we might pull out dandelions and docks. Much of this wild vegetation withers away to ground level after flowering, only to reappear almost miraculously in spring, after the winter rains have recharged the water table.

As climate drifts towards a similar regime we may find that these species, like asphodels and early summer-flowering *Gladiolus* species, will find our gardens increasingly to their liking.

There should also be an edible bonus from the change in weather patterns. In most parts of the British Isles, tomatoes are a glasshouse crop and attempts to grow them outside are fraught with risk. Low summer temperatures can mean that they never ripen properly, leading to a glut of green tomato chutney in autumn.

Some years ago the National Vegetable Research Station carried out a survey to identify areas of Britain where the climate would be suitable for outdoor tomato crops. Their studies showed that in 'normal' years successful crops could be expected in lower lying areas roughly as far north as Oxford, in a few select areas along the South Wales coast and in the pocket of land formed where North Wales butts on to Lancashire.

By the middle of the next century, the outdoor tomato map of Britain will need to be substantially redrawn. In the summer of 1990 I managed to grow a late but passable crop of outdoor tomatoes in County Durham. In fifty years time we should not need to cram them into our glasshouses during the summer; they will be able to take their place beside the cabbages and onions. Eventually sweet peppers and aubergines could also make the transition from glasshouse to open garden, particularly in the parts of southern England that will develop a Mediterranean climate.

At present, attempts at growing sun-loving outdoor crops like sweet corn anywhere north of Watford are all too often an act of folly, but all that is about to change. Many of those alluring, exotic vegetables that beckon from the seed catalogues but always seem to disappoint should at last come into their own. One estimate suggests that, as the climate warms up, the northern limit for growing reliable crops of sweet corn in Britain will advance at the rate of ten kilometres per year.

This wave of warmth means that culinary delights like okra are a possibility (could gumbo become a classic English dish?), while the seasons for many everyday vegetables and fruits ranging from potatoes and French beans to strawberries and blackcurrants will be steadily advanced and extended. The demand for varieties that will perform well at different times of the year, so that advantage can be taken of prolonged growing seasons, will mean that plant breeders will need to work hard to select cultivars that will produce crops over a wide range of daylengths.

For gardeners in the southern half of England, 1989 and 1990 proved to be the years when they really reaped the benefit of planting such exotic fruits as figs, outdoor grapevines and even citrus fruits in pots. Those who were lucky enough to have fairly mature fig trees managed to harvest remarkable crops. It was also a bumper year for glasshouse grapevines, with good fruit set and early ripening; even outdoor vines that usually produced grapes that were only fit for making wine ripened an edible dessert crop.

One of the secrets of successful cultivation of these exotic soft fruits is to choose your varieties carefully. Growing a grapevine from a pip extracted from some particularly tasty variety of imported dessert grape is almost guaranteed to bring disappointment.

I made exactly this mistake and planted a seedling vine in my greenhouse. Needless to say, it is completely confused by our seasons and produces a forest of vegetative growth that threatens to push the roof off the greenhouse. Eventually it grudgingly yields a bunch or two of insipid grapes in October which sometimes ripen by Christmas, by which time they are riddled with *Botrytis* and quickly wither until they look and taste like mummified raisins. I should not have been so mean: if I had spent a few pounds on a reliable named variety I would now be picking bunches of grapes every summer, and I might not have to fight my way through a jungle of branches and tendrils to the potting bench. The moral of the story is that if you plan to attempt to grow these more demanding fruits from warmer climates, buy the best varieties that are available.

One of the greatest feats of gardening one-upmanship in Britain is to grow and harvest a crop of oranges or lemons. A number of gardeners who had bothered to grow named varieties that are tolerably well-adapted to our present summer climate achieved this gardening distinction in 1990. They should savour the horticultural kudos while it lasts, because future generations will probably consider fresh-picked, home-grown oranges on the dinner table as being rather mundane. Such will be the legacy of a changing climate. What is exceptional today will be commonplace by the end of the twenty-first century.

Two other closely-related Mediterranean fruits that will become easier to cultivate as the climate becomes warmer will be peaches and nectarines. Both of these are hardy trees in our present climate, and can be persuaded to crop outside against south-facing walls, but the major problems with these fruits at present are that they flower early in the season, so that their blossom is prone to frost damage. This may eventually become less of a problem if springs become earlier and milder, while plenty of summer sunshine will ensure that any fruit that does set should ripen well.

The early flowering of peaches also means that there are very few pollinators available when the flower buds open. Experienced peach growers hand-pollinate their crops with a paintbrush or rabbit's tail. Milder springs will encourage bees to emerge earlier, as they did in 1989 and 1990, and help in the task of pollination.

Bee activity can make a substantial difference to crop yields, and recent work in Germany has demonstrated that a good supply of bumblebee pollinators will boost strawberry crop yields by about a third. All soft-fruit growers will welcome the beneficial effects that earlier springs and hot summers will have on pollinator activity and fruit yields.

Changing climate patterns will also have a significant effect on crops grown under glass. Higher mean winter temperatures will lower the cost of heating, so out-of-season production of vegetables and raising early transplants will become more profitable. Warm, humid winters would mean, however, that more care will be needed to guard against winter fungal infections of *Botrytis*. Glasshouse hygiene will be paramount.

In summer a conventional greenhouse will be fine for your cactus collection in southern Britain, but it will need thorough ventilation and shading if you plan to keep less drought-tolerant species at their best. Greenhouse designers will need to learn from the experience of gardeners in the Mediterranean, where shade houses are commonly used. These replace glass with fine mesh netting, which maintains temperatures and keeps the wind off the crops, while at the same time preventing sun-scorch of delicate foliage.

When we can move current glasshouse crops, like tomatoes and cucumbers, into the warmer open garden, space will be available to cultivate more exotic plants under glass. Easier conditions for overwintering tender plants will mean that more gardeners might also be tempted to try their luck with tropical orchids and other exotic plants. The space could even be devoted to melons, which should also succeed outside in longer, hotter summers.

Our gardening forefathers laboured long and hard to construct hotbeds of straw and horse manure, which simmered away and provided a warm and moist environment for melon crops. Gilbert White, the eighteenth-century parson who described the natural history of the village of Selborne in such detail, was a passionate melon grower and invested large sums of money in hotbeds so that the table could be graced with home-grown fruits. By the year 2050, home-grown melons may be commonplace in glasshouses, and will probably be a viable outdoor crop in the extreme south of the country by the end of the twenty-first century, provided that gardeners are willing and able to provide them with the vast quantities of water that they will need.

The British public takes an intense interest in gardening. It is because garden plants are under constant scrutiny that they will be such good indicators of climate amelioration. We will notice its effects first in our own back yards.

So, well within a human life span, it seems that future panelists on that hardy radio perennial 'Gardeners' Question Time' will need to bone up on a whole new spectrum of gardening questions. How best to store home-grown citrus fruit, rather than how to avoid frosted begonias, could well become the hot issue by the middle of the twenty-first century.

THREE

A Plague on All Our Crops

> There is something cheering in the knowledge that Nature always hits back. It is metaphysically inspiring, if physically discouraging.
>
> JOHN STEWART COLLIS,
> *The Worm Forgives the Plough*

For many years I have regularly lectured to gardeners and natural history societies on wildlife gardening. At the end of every lecture, as the last slide fades from the screen and the lights go on, I can usually predict the first question from the audience.

'How can I kill slugs?' someone will ask. Despite the fact that I will have spent much of the last hour ardently proclaiming the conservationists' message, the first question is always loaded with murderous intent.

No animal is so universally loathed by people who cherish plants. You can persuade them to tolerate woodlice and spiders or even millipedes, but I have yet to meet a gardener of any description who welcomes the sight of a slug anywhere within his or her plot. Even wildlife gardeners dislike them, for they will chomp their way through a batch of carefully nurtured wild flower seedlings with the same alacrity that they display in felling succulent delphinium shoots in spring.

Wherever gardeners gather together, the conversation eventually turns to slugs and how to dispose of them.

The recent run of mild, wet winters have been very kind to the slug population and their numbers have multiplied alarmingly. The general trend in precipitation throughout the

1980s was towards a shift in rainfall patterns, with more rain falling in winter and less in summer. This trend seems set to continue according to many climatologists, so by the middle of the next century western and northern parts of the country can expect milder winters with substantially more rain.

If slugs are caught out in the open in dry, sunny weather, their chances of survival are slim, but they are very adept at finding small, damp crevices where they can safely sweat out the daylight hours in the hottest months of the year. Slugs often migrate downwards in the soil to the deeper, moister layers, so one of the best control methods in summer is to cultivate the soil on a hot, sunny day, bringing them and their glistening white eggs to the surface; birds and desiccation will finish off the survivors.

Slugs and snails are active even during a drought, because they do most of their feeding at night. If you look around your garden in daytime during the kind of parched conditions that we experienced in 1989 and 1990, you might well believe that you do not have a slug problem. Try going out again with a torch at around eleven o'clock on a warm, humid night in mid-summer. The sight that will greet you in your herbaceous borders will fill you with horror. The vegetation will be crawling with slugs slithering over leaves and flowers in search of a meal. You may well have trouble sleeping afterwards.

Snails are even better than slugs at surviving drought. Some species have a horny lid, called an operculum, which they can use to seal their shells when they retreat inside; most secrete a film of mucus over the entrance, which dries to form an effective barrier. In the Mediterranean countries, snails use this technique to shut up shop in the hottest season, entering a period of summer dormancy called aestivation. In late August, the tinder-dry vegetation of the Greek Islands is often covered with aestivating snails, safely sealed inside their shells until the winter rains return. In Britain, garden snails often employ a shorter-term version of this technique, usually choosing to crawl into the centre of plants like pampas grass, which are particularly safe hideouts.

A PLAGUE ON ALL OUR CROPS

So it seems certain that slugs and snails will tolerate and probably profit from whatever our climate delivers in the next few decades. We gardeners are faced with a major mollusc problem, as are farmers, whose winter-sown crops are often very badly damaged. It has stimulated intensive research aimed at finding an environment-friendly molluscicide.

The best solution might seem to be to identify and nurture some of the slug's natural enemies, but there are not many effective candidates. Hedgehogs are undoubtedly voracious mollusc predators, and there can be few sounds as tuneful to a gardener's ear as the snuffles and muffled crunching of a hedgehog eating snails and slugs in the twilight hours in the flower borders.

But sadly the hedgehog is far from being domesticated, and even in large gardens they are reluctant to take up residence for long. They seem to be natural wanderers, which probably accounts for the enormous number of them that end up as road casualties. The only sure-fire way of coaxing them to stay seems to put out food – not bread and milk, which is bad for their digestive systems, but cat food. Unfortunately, if you employ this tactic you are likely to be saddled with a freeloading hedgehog with little remaining appetite for slugs, and will also have attracted every cat in the neighbourhood to your back door. And most gardeners of my acquaintance dislike cats almost as much as slugs. If only cats ate slugs!

There is another, smaller slug predator that shows some promise. The Henry Doubleday Research Association, which promotes organic methods, is developing new techniques for pest control. It is supporting research at the University of Wales in Cardiff, where William Symondson is trying to find ways of exploiting one of the common field slug's natural enemies, a ground beetle called *Abax*. He is trying to persuade Abax to breed in large numbers but the day when we can buy these beetles in a garden centre is still a long way off.

Even amongst the most environmentally-orientated gardeners there is a strong temptation to resort to slug pellets when all else fails. But a chemical minefield of metaldehyde sown in the borders is not particularly effective. Slugs may only have a brain the size of a pinhead but when it comes to poisons they are not stupid. The problem with current chemical control methods is that they apparently have a taste that slugs do not like. This means that they eat a small amount of the poison, which may be enough temporarily to paralyse them, but they then shy away without taking a fatal dose. Many subsequently recover and crawl away to safety, and some of the fatalities that do occur are probably due to the fact that the slugs are paralysed and caught out in the open as the sun comes up, so that they die of desiccation rather than poison.

So the search is now on for a palatable poison for molluscs. There are some indications that success may not be too far away.

Tests at the world-famous Rothamsted Experimental Station in Hertfordshire provide a glimmer of hope. The most promising alternatives to metaldehyde-based pellets, which are potentially dangerous to pets and wildlife, are aluminium salts. These kill slugs on contact under laboratory conditions, and are already marketed as a product which is sprayed on to the soil surface with a watering can. The drawback with these soluble products is that they need to be applied to soil in enormous doses if they are to be really effective in the field. And molluscs are even more reluctant to eat bait pellets which contain the metal salts than they are to feed on metaldehyde.

But there is at least one aluminium compound that slugs actually seem to like. There are encouraging signs that new baits containing a substance called aluminium acetylacetonate may turn out to be far more effective than current products. The new formulation has scored well amongst gourmet molluscs, leaving a trail of dead snails that seem to have died with a pleasant taste in their mouths. It is certainly more effective than other soil-acting, contact slug poisons, like the aluminium sulphate which is currently in use. If it proves to be less damaging to wildlife than present molluscicides, it may have a promising future which, perhaps, might be more than can be said for the slugs.

But it seems more than likely that slugs and snails will remain an intractable horticultural problem which will get worse as the climate changes. The same is also true of that other loathsome garden pest, the greenfly.

Populations of insects like greenfly, capable of several generations per season, should respond rapidly to climatic changes as they occur. They seem certain to benefit from a rise in average temperatures.

Crop pests in general, opportunists *par excellence*, should thrive on warm winters. It is no coincidence that 1989 was one of the worst years on record for greenfly infestation of crop plants and that 1990 was not much better. They survived the mild winters of 1988/89 and 89/90 very successfully, and a warm, early spring allowed them to disperse quickly and colonize garden plants and agricultural crops like winter wheat. In the early summer of 1989 there were fears that stocks of suitable agricultural insecticides might run low.

Aphids do great damage by sucking the sap of plants, but their indirect effects are even more devastating. Their mouthparts are like tiny hypodermic syringes, beautifully designed for penetrating between the cell walls of a leaf and finding the sugary sap that courses through a plant's veins. As they feed they pick up virus particles and their mouthparts become infected needles, transporting and injecting viruses into new hosts.

Part of the key to the aphid's success lies in its adaptable sex life. Under favourable conditions, female greenfly can conveniently switch from the slow process of egg laying to giving birth to live young, which they produce at a phenomenal rate. This allows a well-adapted greenfly to saturate a host plant with perfect copies of itself. Within two weeks of birth, these clonal offspring start producing young of their own, so consistently high summer temperatures and high winter survival inevitably lead to greenfly population explosions.

At the end of a growing season, aphids revert to conventional sex, mating and laying eggs that are resistant to environmental stresses. The genetic variation between the offspring, which is an inevitable consequence of sexual reproduction, means that a wide range of genetically distinct individuals are produced in winter. This acts as a form of inherited insurance policy, generating new types of individual which may well do better in a changing environment.

So during periods of climatic stability, the clonal individuals can capitalize on a favourable environment; when the environment is less predictable, the sexual individuals can be relied on to come up with genetic novelties that will guarantee the future of the species. No animal has a better survival strategy for dealing with the seasonal confusion that we can expect in the next half century. Greenfly can enter the era of climate instability with supreme confidence.

If winters run their recent mild course, we can expect similar patterns of greenfly problems every summer. The single saving grace is that this potential pest explosion may well be balanced by better survival of greenfly predators. The agrochemical industry may see the dawn of a warmer climate through the rosy glow of increased pesticide sales; the environmentalist's scenario is that the natural system of checks and balances will prevail.

'The balance of nature' has become something of a naturalist's cliché, but it is a well tried mechanism that really does work. Take the case of root fly versus cabbage aphids, for example. Scientists studying the cabbage root fly at the Institute of Horticultural Research at Wellesbourne in Warwickshire have shown that cauliflowers that carry a moderate infestation of cabbage aphids are immune to cabbage root flies. The presence of the aphids seems to unsettle the flies, which then decide to breed elsewhere. Earlier studies have also shown a similar antagonistic interaction between the aphids and some moth pests of cabbages, whose droppings contain a compound called sinapic acid, which cabbage aphids dislike intensely.

If you must get rid of aphids, the best strategy may be to leave it to the professional aphid predators, like ladybirds, whenever possible.

Recent mild winters have been kind to ladybirds, too. In the winter months of 1989/90 gardeners all over Britain noticed large numbers of this efficient greenfly destroyer emerging from hibernation on mild days. These beetles reduce populations of overwintering aphids and help to control their spread in spring. High winter survival of ladybirds can lead to the kind of 'ladybird plague' that we saw during the hot summer of 1976, when vast, hungry swarms of these colourful beetles devoured the aphid population and then roamed the countryside in search of food.

During the late summer of 1976 I was on holiday in Cleethorpes when the ladybird plague arrived. I vividly recall the sight of bathers feverishly brushing red and black blobs from their exposed flesh or picking the struggling insects out of their ice creams, while shopkeepers swept up shovel-fulls of the insects and summarily disposed of them. The entire length of the promenade railings was coated with clusters of scarlet and black beetles, resting in the sun before they moved on in search of food.

Dedicated ladybird watchers, like Dr Mike Majerus at Cambridge University, predicted another massive population explosion of seven-spot ladybirds in 1990, but this time it failed to materialize. The mid-summer drought might have been partly to blame, causing a reduction in the aphid population, but the main cause seems to be an exceptionally nasty little fly parasite which quite literally eats ladybirds alive. The adult *Phalacrotophora* fly lays its eggs on ladybird pupae, which the hatched maggot slowly consumes. Dr Majerus reported that as many as 80 percent of ladybirds had fallen prey to this insidious parasite in 1990.

This unexpected setback for ladybirds, during a period when climate was very much to their liking, illustrates one of the major problems in predicting the effects that climate will have on pests and predators. This relationship between the hunters and the hunted is rarely a simple one, and the hunters themselves may well fall prey to predators. Fortunately these feeding relationships are complex, and there are several other species, like hover flies and lacewings, that can be relied on to help to keep greenfly in check.

Adult hover flies are for the most part pollen feeders, although their football-jersey-striped abdomens give them a passing resemblance to carnivorous wasps. The larvae, on the other hand, feed entirely on aphids, downing as many as two hundred per day. They are a singularly unattractive stage in the hover fly life cycle, resembling a cross between a miniature leech and maggot, but they wreak havoc amongst the greenfly hordes.

If you watch a hover fly larva wriggling its way into a dense aphid colony, the effect on its prey is electrifying. Panic spreads rapidly as the aphids emit alarm pheromones – volatile compounds that signal to the rest of the colony that they are under attack. Those that are quick enough can withdraw their needle-sharp mouthparts and make their escape. Some will simply drop off the plant. But for the unlucky ones, the hover fly larva's jaws promise a violent death. It takes about thirty seconds to despatch each victim.

Hover fly numbers do seem to increase in years when there is an abundance of aphid prey available, and occasionally their numbers can soar dramatically. In early August 1975, South Humberside was plagued by massive swarms of hover flies, following a period when greenfly numbers had been very high on cereal crops. During the hover fly invasion Grimsby Environmental Health Department was forced to issue numerous reassurances to householders who mistook the insects for stinging wasps.

In this increasingly environment-friendly age most gardeners know that they can increase hover fly numbers by avoiding the use of insecticide sprays and providing a plentiful source of pollen for the adult insects. Members of the daisy family, in the form of *Gaillardia*, *Achillea*, *Calendula* and *Coreopsis*, to name but a few genera, are particularly good pollen sources for these useful predators. A garden with a flower border that is well stocked with pollen-rich ornamentals has a head start in the fight against aphids.

As in the case of slugs, active research is going on into ways to combat greenfly using natural predators. One promising line of defence for agricultural crops may be to adopt the organic gardener's strategy and plant pollen sources around the edge of cereal fields.

Organic gardeners will be familiar with the lavender blue flowers of *Phacelia*, which exerts a magnetic attraction on hover flies. It is an American ornamental which is also sometimes grown as a green manure. One of its characteristics is that it has a long flowering period, providing a constant supply of pollen. At Southampton University, Dr Steve Wratten has been engaged in planting this colourful species around the edge of cereal fields and monitoring its value for the predatory hover fly population.

Phacelia pollen has a particularly characteristic shape, so Wratten and his colleagues have been able to demonstrate the valuable part played by the flowers in boosting hover fly numbers by looking for the pollen in the insect's gut. It may be some time before flower borders become a familiar sight in cereal fields, but this is certainly a promising step towards natural control of insect pests. There does, however, seem to be a certain irony in the fact that modern agricultural practices have done so much to wipe out wild flowers and the useful native insects that they support, and that an imported American species is being used to try to redress the balance.

Before leaving the subject of greenfly plagues, I cannot resist mention of one more of the aphid's enemies, the ferocious lacewing fly. This is a deceptively fragile-looking insect, green, the size of a medium-size moth, with weak flight, bulging golden eyes and wings that appear to be made of the finest gauze. It often comes into houses through lighted windows as the days shorten in late summer. It is an aphid assassin on a grand scale.

Cherish the lacewing fly. It lays its delicate egg on a long stalk on the underside of hawthorn leaves. The larva that emerges is supremely ugly and would merit a part in any Hollywood horror movie. It sinks its pincer jaws into aphids and sucks them dry, tossing the empty skins over its shoulder so that they become impaled on the spiky hairs on its back. Should you notice a small heap of ladybird corpses meandering around on the surface of a leaf, in all probability it will be a lacewing carrying a portable mortuary as camouflage.

Canadian researchers are already introducing lacewings into orchards on a large scale. We must hope that any climate change that favours the ladybird will do likewise for the lacewing.

The red spider mite and the glasshouse whitefly are two notorious pests that are guaranteed to send a shiver down the spine of any gardener. Both bugs will benefit from higher seasonal temperatures.

The red spider mite is barely visible and it is usually the symptoms of its feeding that arouse suspicion that something is amiss. The first sign of spider mite activity is often a pale yellow blotching of leaves which soon develops into a bronzing. By the time this stage is reached, the infestation is very difficult to control. The pestilence rapidly proceeds to the stage where the leaves and shoots are festooned with a web, with spider mites running up and down the threads like sailors swarming up the rigging of a sailing ship. This was a depressingly familiar sight for gardeners in the hot summers of 1989 and 1990, when red spider infestations were particularly severe.

Spider mites can breed at a prodigious rate, and the higher the temperature, the faster they multiply. In summer in a glasshouse they can develop from egg to adult in about eight days, each female producing about a hundred eggs. In winter they hibernate, and no matter how efficiently glasshouse hygiene is maintained, a few always seem to survive into spring. They can crawl into inaccessible crevices, like electrical sockets and hollow canes, where fumigants do not reach.

They thrive in warm, dry climates and during hot summers will move out of glasshouses and infest garden plants. By the middle of the next century the glasshouse spider mite will probably be a general garden pest, attacking everything from roses to violets.

These mites are extraordinarily difficult to control. Their rapid reproduction rate has allowed them to evolve resistance to even the most toxic chemical sprays. The only practical organic control methods available are to spray plants with a fine mist of cold water – they dislike high humidity and this will deter them – or to use another Chilean mite which preys on them as a means of biological control, using the principle of setting a thief to catch a thief. These predators are already widely used in commercial glasshouses, but if this pest spreads out into the general garden, this 'natural' form of control will be extremely difficult.

The prognosis for glasshouse whitefly is equally worrying. These too have already succeeded in moving out of the glasshouse and on to general garden plants during hot summers. As a subtropical species which was accidentally introduced into Europe, it has, until now, done most damage in the sheltered environment of greenhouses. Apart from the direct damage that it inflicts, its most unpleasant symptom is that it covers leaves with sticky honeydew, which becomes a fertile breeding ground for fungal infections like sooty mould. Fuchsias, tomatoes and cucumbers are especially vulnerable, but the tiny white flies have catholic tastes and will found colonies on a wide range of species.

As with the red spider mite, the whitefly's life cycle is accelerated by rising temperatures, and its eggs can develop into adults within three weeks once the mercury thread creeps above the 20°C mark. Once hatched, each female lays about two hundred eggs. Any disturbance of a badly infested crop releases a miniature snowstorm of microscopic flies.

Cold winters kill outdoor whitefly, but if frosts become rarer more individuals will successfully overwinter and found new colonies in spring. Like red spiders, whitefly have become resistant to insecticides and those that kill adults have little effect on the eggs or scale-like nymphs. One of the best available control methods is a parasitic wasp called *Encarsia* which lays its eggs in the whitefly scales, but although it controls the pest well in the warm confines of a glasshouse, its effectiveness outside, in the variable climate of the garden, is unknown.

One Russian research group has refined an organic method of whitefly control which may offer a means of protecting crops on a larger scale. Sticky yellow strips are already available which are known to lure whitefly to a tacky death, but the Russians have injected a further element of high technology into these by studying the eyesight of the pest. They have tested dozens of different shades of yellow and have now come up with a sticky trap of a hue which is reputed to be irresistible to whitefly. If these insects do become major garden pests as the temperature rises, we may need to deck our gardens with hundreds of these yellow flypapers.

The thought that there may be dozens of new pests that will take advantage of a kinder climate is a disturbing one. Another aspect of the pest problem is that plants that are drought stressed during long dry summers may be more vulnerable: most pathogens attack weakened plants preferentially. There are many such imponderables waiting to manifest themselves as summers become hotter and winters lose their chill.

This change in climate might also result in some rare pests spreading their home range. During the red-hot summer of 1990, many gardeners in Berkshire reported infestations of small red beetles which quickly reduced leaves and stems of lilies to mangled wreckage. The culprit, the lily beetle, is at present confined to Surrey and Berkshire and a few other localities in the south. My guess is that the lily beetle – and many other garden pests that are currently rare and confined to the warmer areas of the country – will become established in other localities as the climate changes. The lily beetle is a beautiful insect, of the most striking scarlet. Unless, of course, you happen to be fond of lilies, in which case it is the ugliest creature on Earth.

FOUR

A New Gardeners' Calendar

> The average gardener, who we must conceive of as being a lazy, pleasure-seeking so-and-so, has traditionally waited upon the Easter holiday before getting down to the annual tasks in the garden that are associated with the dormant season, with catching up and pressing on. For form's sake he cuts down the herbaceous stuff in the autumn, but after that it's five months of feet up while the weather does its worst.
>
> CHRISTOPHER LLOYD, *The Adventurous Gardener*

Well before the middle of the next century, it seems certain that the gardening calender will need to be substantially rewritten. Winter, traditionally the season to put your feet up in front of a roaring fire and thumb through next year's seed catalogues, with a Scotch within easy reach, will be a busy time. Forget the advice about greasing the lawn mower before you lay it up for the winter. You are going to need it in January.

One of the effects of climate amelioration predicted by the Meteorological Office is that a temperature rise of about 1°C will reduce the frequency of January frosts by over 25 percent. Winters look like becoming wetter and warmer.

Most temperate plants grow best at temperatures between about 10°C and 35°C. Small fluctuations within this range may hardly seem significant to us, but they can make a dramatic difference to plant biochemical processes, like photosynthesis. Constant small changes have a cumulative effect. A one percent rise may not sound much, but its effects on annual growth will be clearly visible.

In gardening terms it means that the length of the growing

season will increase considerably. The upshot is that the leafy suburbs will echo with the sound of hovermowers during the short daylight hours of winter weekends.

It will also be important to keep the hedgetrimmer handy: that rampant *xCupressocyparis leylandii* hedge, which the neighbours complain about in summer, will probably grow another couple of feet in the winter months too.

Milder winters will mean switching some gardening operations from spring to autumn. With less risk of frost, especially in the southern half of the country, it will be safer to prune roses in October to take advantage of winter growth. By the same token, late autumn will be a good time to plant almost all hardy species, to allow transplants to establish before summer droughts strike.

Some old gardening lore may also be due for revision. Although autumn pruning of established shrubs and roses will be safer, pruning newly planted bushes at that time of year will probably be something to avoid. There is evidence that root and shoot growth are often linked, so pruning shoots can actually restrict roots. Bonsai growers are familiar with the reverse case of this phenomenon, in an extreme form: constant removal of roots is part of the essential technique for producing bonsai trees.

Television gardening pundits take great delight in cutting back perfectly healthy shoots of newly transplanted bushes virtually to ground level. Spending several pounds on a vigorously growing plant, carefully chosen from a garden centre, and then hacking away all the best bits and consigning them to the compost heap or bonfire has always seemed to me to be a savage way to go about gardening.

Some experts even recommend pruning roots before planting. The logic behind this excess of zeal with the secateurs is that root pruning is supposed to induce vigorous new root formation, while shoot pruning reduces leaf surface area, so the plants lose less water while they are becoming established.

Recent research shows that, like many pieces of traditional

gardening wisdom, this advice needs to be applied with caution. Root pruning does stimulate new growth, but this merely compensates for the roots that have been removed. Shoot pruning may reduce water loss and help the establishment of trees and shrubs in dry conditions, but it also reduces the leaf area available for photosynthesis and wastes all the stored starch in the pruned stems that would have contributed to vigorous growth in spring. Pruning roots and shoots simultaneously is a recipe for total disaster. So beware of experts wielding secateurs; plant your plants intact, in autumn when there should be plenty of water from winter rain. The only additional treatment needed will be to attach them to stout stakes to prevent winter gales from rocking them out of the ground.

If the pattern of dry springs we have experienced lately were to become the norm, autumn sowings of annuals would be essential to ensure good germination and to allow them to produce early, deep roots that would help to cope with water deficits in the following year. This would also have the effect of inducing them to flower earlier, so some of the annuals that we now associate with high summer will begin to flower in late spring, exaggerating the 'June gap' when the flower garden is relatively bare.

Paradoxically, in a warmer Britain vegetable breeders will need to produce varieties that are more tolerant of low temperatures. This is because winter sowing of vegetables, as is currently practised with some Japanese varieties of onions, would become a recommended precaution for avoiding the worst effects of summer droughts.

So autumn and winter will be busy seasons. There will be much to do in the short daylight hours of December, which will make life difficult for avid gardeners with nine-to-five jobs. Dark evenings will mean that most work will need to be done at weekends.

Warm, wet winters will do no favours for some of the plants that we traditionally associate with frost and snow. Alpines overwinter best in cold, dry conditions; waterlogged soils cause them to rot at ground level. Anyone who has unsuccessfully attempted to keep lewisias alive through mild, wet winters will be painfully aware of the challenge that wet, warm soil poses to plants from high, frozen places. Alpines that produce a neat cushion of foliage, like saxifrages, *Silene acaulis*, dionysias and *Androsace* soon lose their characteristic compact growth forms if winter temperatures favour continuous growth. For these species, perfect drainage will become even more important. Growing alpines well in the southern half of the country will be a severe test of horticultural skill.

A warm winter will promote growth of garden plants but, as always, it will be adaptable weeds that will really cash in on a kinder climate. Species like chickweed and groundsel which can already flower in mid-winter will flower and seed prolifically right through the depths of December and January.

Groundsel is a prime example of the kind of supremely adaptable weed that will exploit the situation. It flowers and sets seeds when the days are darkest, coldest and shortest, succeeding in producing winter seed which germinates in spring to produce vigorous plants that take advantage of abundant soil moisture to grow into lush giants that spawn thousands of feathery seeds. When these are shed they contribute to a further generation that can grow in the parched conditions of a summer drought, even in the cracks between paving stones, perhaps only growing an inch or two high but still successfully producing seeds. For gardeners who want to break the cycle of multiplication and exert some control on adaptable weeds like this, it will be essential to do the weeding thoroughly in winter, eliminating the plants that will produce the prolific spring generation.

This tendency of plants to continue growth over mild winters will create many problems. In recent winters garden plants have continued to make soft growth, providing a 'green bridge' between autumn and spring, allowing fungal and insect pests to overwinter in unusually large numbers. In future, clearing away dead stems and faded flowers in autumn will become even more important, to reduce the risk of them harbouring next year's fungal diseases. Late winter use of insecticides, to knock out latent aphid infestations, may be *de rigueur* amongst those who still favour chemical control of pests.

Aside from killing pests, heavy frosts play several other important positive roles in the garden which are often overlooked. Seeds of many primulas, violets, hellebores and alpines all need prolonged cold treatments before they will germinate, so in a frost-free climate the domestic fridge will act as a substitute. Hellebores need two seasons of winter chill to coax them into growth, and seeds of roses and many shrubs share this requirement for dormancy-breaking stratification. Will future winters be up to the task of freezing these seeds into life? In October, we had better be prepared to clear the shelf just below the freezer compartment of the refrigerator for flower pots.

When water freezes it expands, exerting enormous forces on anything that happens to be in the way. Householders who have had to call out a plumber to repair burst pipes after going away for Christmas and leaving the heating off will recognize the phenomenon. Alternating cycles of freezing and thawing of water held between soil particles forces them apart, so that lumps of clay crumble into a fine tilth in spring. This service comes free at present, courtesy of the infamous British winter, but we may expect to have to dig more thoroughly in a frost-free climate to achieve the same effect.

Some climatologists predict that certain parts of the country, especially the north and west, can expect as much as 20 percent more rainfall in winter in the next century. One unfortunate aspect of this increased precipitation will be that it will wash away soil nutrients. Building up soil fertility is a key principle of good gardening, but if land is left bare over winter, hard--won soil fertility can disappear down the drain with the first heavy downpour. To be available to plants, soil nutrients must be soluble in water; if they are soluble in water, they will be washed away by rain.

One way to break this vicious circle is to use green manures, ground-cover crops which absorb nutrients and temporarily lock them up, out of the reach of rain, until the plants are dug back into the soil in spring. Green manures are all fast-growing species that are sown, allowed to grow and then dug in. They decompose in the soil, releasing nutrients and building up humus levels. The increase in organic matter is particularly valuable in sandy soils, especially during dry summers, because it increases the water-retaining capacity of the soil.

There is a wide choice of green manures. One of the best is *Phacelia tanacetifolia*, not least because it is easy to dig into the soil. The masses of powder-blue flowers are highly decorative and attract large numbers of hover flies and bees. In some areas of Britain its winter hardiness is somewhat dubious at present, but climate amelioration should take care of that.

Crimson clover is an alternative and very attractive species, which, like *Phacelia*, may be marginally hardy at present but has certainly survived mild winters in recent years. This produces long, crimson flower heads which attract a constant stream of bumblebees. Crimson clover has the additional advantage of bacterial root nodules, which convert gaseous nitrogen in the air into nitrate fertilizers, which are released into the soil when the crop is dug in.

There are several other green manures that fix nitrogen in this way, including alfalfa and winter field beans (a small-seeded variety of the broad bean), red clover, winter tares and trefoil, all of which are winter hardy and particularly suitable to growing in plots that would otherwise lie idle over the winter.

For a really fast-growing green manure that will increase soil humus quickly there are few plants better than grazing rye, which has a deep root system that will improve the soil structure. It also suppresses weeds very effectively over the winter months. If it has a drawback, it is that it is very hard work to dig this crop back into the soil.

So, by the time winter is over and you have finished cutting the lawn and digging in the green manure, have pruned the roses, trimmed the hedge, fretted over the rotting alpines, finished the weeding and lean exhausted on your spade, having completed the extra digging needed to make up for the lack of frost action on lumps of clay, you will probably be ready for a rest. But you will barely have time to draw breath. No sooner will the clocks go forward than spring will arrive in a flash.

If current trends are any indication, future springs could be earlier and shorter and they will be times of great anxiety. This change in the pace of seasons will have a significant effect on many gardening events, including flower shows, which will need to be retimed. Recent mild winters have made life difficult for Britain's competitive growers of daffodils, alpines and primulas whose prize plants have peaked too early.

Too much warmth too soon confuses certain plants. Primulas are particularly prone to reckless flowering, and polyanthus and auriculas have a habit of producing early flowers that quickly become tatty in the unsettled weather that leads up to spring. This is not too serious a problem in these species, except for gardeners who are trying to produce prize specimens, because, by literally nipping these precocious buds in the bud, flowering can be postponed until the weather pattern is more secure. However, for many garden plants an early commitment to flowering is irreversible and can be very destructive.

When mild days in March loosen the flower bud scales on fruit trees and accelerate the growth of plants that have never really stopped growing throughout the winter, our gardens will be at greatest risk. The hazard comes from sudden, vicious frosts that may linger until the end of April. Eventually, when climate change has progressed to the point where much of the southern half of England may be relatively frost-free in spring, we will be able to breathe easy again, but in that tricky transitory period, where winters are warm but late frosts are not yet extinct, there will be many losses. Species like *Clematis montana*, which responds to a mild winter by initiating flower buds in the ridiculously early month of February, will pay for their eagerness with a damaged crop of flowers. Unlike primulas, *Clematis montana* does not seem to be able to compensate for losses of early buds by producing replacements.

As soils warm up quickly in spring there will be a temptation to sow early, but cloches will be a wise precaution. It will be a time for anxiously watching the weather forecasts for frost warnings, ready to take whatever steps are possible to prevent early lettuces from succumbing to occasional cold snaps.

Here there will be opportunities for manufacturers to market useful products that will provide temporary cover for tender crops. Floating mulches, like light agryl sheeting, which has the consistency of a nappy liner and can cover the crop in an almost weightless insulating blanket, will be a boon; ideal for unfurling over plants at risk when the weathermen issue their warnings.

It's July. Already no rain has fallen for six weeks and there seems little prospect of any arriving in the foreseeable future. If there will be a time to take a breather in the new gardening calendar, this will be it. There will be no point in planting vegetables or flowers during this period unless you have plenty of water available and the time and resolve to administer it. So the best course of action will be to get out the deck chair, extract a chilled bottle of wine from the refrigerator and enjoy the fruits of your labours.

Well, perhaps. There will still be plenty of jobs that you could be getting on with.

'Mulch' will be a key word in the vocabulary of the twenty-first-century gardener. Mulching, which entails covering the soil surface with a deep layer of peat, bark, compost, or similar organic material, will help to solve the drought problem by conserving soil moisture, something that will become a constant chore, especially in the predicted Mediterranean climate of the Home Counties.

In preparation for hot summers, the prudent gardener will do well to master the black art of composting. All those prunings that currently end up on the local council tip will suddenly become a valuable commodity, converted to an instant mulch or compostable material with the aid of a shredder.

The current trend towards recycling wastes will mean that many more commercial mulches will appear on the market, replacing diminishing peat resources. Some local highway authorities and parks departments in Britain already use shredders to recycle prunings. In New York State, for just a dollar, it is possible to have your Christmas tree recycled through a shredder after the festivities finish and returned to you as mulch in a bag, for reuse in the garden. Much of what we throw away today will become valuable mulching material in a drier climate.

In November 1989 Chris Patten, then Secretary of State for the Environment, set a target for the recycling of half the recoverable material in Britain's dustbins by the year 2000. Whether this ambition will be realized is debatable, but there are several schemes in Britain for composting municipal waste, sewage sludge and waste products from the food industry.

All of these recycling schemes have been prompted by the rise tide of garbage and the shortage of holes to dump it in. What may tip the balance in their favour, turning them into profitable horticultural enterprises, is the need to replace horticultural peat with a cheap, environment-friendly alternative. Peat is a finite and fast-disappearing natural resource, and peat bogs are one of many wildlife habitats that are under intense pressure from commercial exploitation.

So mulching will be an important summer task in the garden. A mulch made from sewage sludge may not have the cachet of finest Irish sphagnum peat, but few gardeners who cultivate plots in the south can afford to turn their noses up at the possibility of using it. Think of the alternative. All those heavy cans of bath water.

Fortunately for British gardeners, generations of scientists in countries with a Mediterranean climate have been grappling with the kind of summers that we have recently experienced, and are now beginning to come up with some effective treatments for controlling water loss from leaves. A changing climate will, of course, provide a vast range of marketing opportunities for the horticultural industry and it is certain that current research into methods for relieving drought stress in plants will eventually result in novel products on the garden centre shelves. Alongside herbicides, pesticides and fertilizers, we will probably see a new section labelled anti-transpirants.

These are compounds that reduce water loss through the leaves of plants. Some work by triggering closure of the tiny pores, or stomata, on the leaf surface, which help regulate water loss. Lichens, which often grow on sunbaked roofs and rocks, are an unlikely but promising source of one of these drought-defeating compounds, and French scientists have carried out successful tests with a compound called usnic acid which is extracted from these primitive plants. Trials show that it can drastically reduce plants' water demand.

Researchers in Israel, where water is always a scarce resource, are masters of growing crops in arid environments and have successfully tested methods for coating plants with water-conserving films of waxes, silicon and even plastic polymers. The Chinese have been testing a similar system for over ten years. The drought tolerance that these treatments confer is substantial and the films are particularly effective in improving seedling establishment after transplanting in dry weather. Perhaps waxing the leaves of our plants will become almost as important as watering them.

The end of a long, hard gardening year approaches and the autumn leaves begin to change colour, though not perhaps with quite the same sudden, flaming spontaneity as they once did. The shortening days will inform plants that winter is on the way, but with no hard frosts to trigger the sudden death of leaves, autumn will be a mellower, more gradual affair. But it will be important to collect those leaves and compost them as they float down. With the prospect of a warm, wet winter in store, a layer of rotting leaves will create the perfect conditions for sheltering slugs and fungal pathogens in the flower border. Rusts, smuts, cankers, mildews, moulds, leaf curls, diebacks, scabs and bud blasts — all are fungal visitors with designs on your plants that are likely to increase in milder winters.

One fungal disease should diminish. Snow moulds are fungi which do most damage in the coldest months of the year, causing brown, slimy, circular patches on lawns of fine-leaved grasses. As their name suggests, they have a remarkable ability to thrive in cold conditions, and specialize in growing under a layer of snow, so that the damage that they do is only revealed when a thaw arrives. As snowfalls in southern England become steadily more feeble, the snow moulds at least should go into a decline. No doubt there will be some equally unpleasant garden pestilence waiting to take their place.

Another exhausting, exhilarating, frustrating but intensely rewarding gardening year has finished. Well, that is not quite true. In the new climate the gardening year will never finish. The lawn and the hedge will still be growing and it will soon be time to begin planting again for next year. But why not put your feet up in front of that roaring fire, pour yourself that malt whisky or a glass of port and thumb through the new batch of seed catalogues. We must preserve something of the old gardeners' calendar, dammit.

FIVE

Sex, Alpines and Aliens

> There is, I suppose, a danger that one might lose one's heart to the Alpine flora, and turn one's back for ever on the dowdy mist- and rain-washed hills of Highland Britain. I can only say that this has not happened to me; spring gentian in Teesdale or the drooping saxifrage in Glencoe still provide a thrill quite undiminished by the fact that I have seen spring gentians by the thousands in the Alps or walked over and ignored drooping saxifrage in Scandinavia. Familiarity with the British mountain flora need never breed contempt.
>
> MAX WALTERS, *Mountain flowers*

The sex life of some of Britain's mosses and liverworts may be an early casualty of climate amelioration.

Apart from raking moss out of the lawn once a year, few of us take much notice of these insignificant plants. A pity, because if we watch them closely, we may see some of the first clear symptoms of climate change.

Imagine walking through woodland in spring, with the first buds bursting to provide a tinge of greenery in the branches. An earthy smell rises from the damp woodland floor, released by a shower of rain. It is, perhaps, a slightly alarming thought that countless millions of microscopic liverwort and moss spermatozoids are frantically swimming through the surface film of water on the green cushions at our feet, in a race to find a female plant and fertilize an egg cell before the sun dries the soil surface.

Liverworts are the most primitive of all land plants, the

closest living relatives of ancestral algae that forsook the primaeval sea and colonized land over four hundred million years ago.

For plants, sex under water is the safest form of reproduction, allowing millions of swimming male spermatozoids to search out female egg cells in liquid security. Once plants conquered land, a spermatozoid's quest for a female counterpart became infinitely more hazardous. It depended on the presence of a surface film of water, provided by rain. For liverwort, moss and even fern spermatozoids, no rain means no sex. Such are the perils of a changing climate.

At Durham University I teach students the rudiments of the unfashionable but fascinating science of plant evolution, which gives me a valid excuse to spend several enjoyable days every spring collecting samples of liverworts and mosses for practical classes. It was as a result of these regular collecting trips that I first began to gain an insight into what climate amelioration might mean to our native flora.

Liverworts are of two varieties, thalloid and leafy. Thalloid types resemble flat, green, fleshy lobes that cling to the muddy sides of ditches, bearing a passing resemblance to miniature green liver. The old herbalists who adhered to the Doctrine of Signatures, whereby any plant that looked like an organ of the human body was thought to cure its ailments, swore by them as a cure for liver disorders. Though their pragmatic brand of medicine may not have been much good, they probably knew a good deal more about these unprepossessing plants than most of us do today.

At a glance, leafy liverworts could easily be mistaken for fragile mosses, with two rows of frail, translucent leaves. Neither type has roots; both have little resistance to drought. They absorb essential water over their whole surface, and lose it with equal ease.

It was after the summer of 1989, at that time the hottest ever recorded, that I began to notice the disappearance of these primitive plants from sites where I had collected them for years. By the spring of 1990 it was clear that leafy liverworts in particular had simply expired in the drought of the previous summer. Thalloid liverworts had fared slightly better, since they are seldom found far from streamsides, but they too had declined.

While this damage to the plants themselves was the result of summer heat, it was a change in rainfall pattern that did for their sex lives. One of my favourite leafy liverworts is a tiny jewel of a plant called *Lophocolea cuspidata*. It forms exquisite, circular colonies of an almost luminous lime green on old conifer stumps in shady woodland. In early spring, after its spermatozoids have completed their perilous journey, it sprouts spore capsules. These appear almost overnight, like forests of slender matchsticks with glass stalks and jet black heads.

We had an exceptionally dry spring in 1988 and 1989, with a bone-dry soil surface, so *Lophocolea*'s spermatozoids never reached their destination at most of my collecting sites. As a result, spore capsules were in short supply.

Mosses too were seriously damaged in the drought of 1989, but many of these are better adapted to surviving long periods with little water. Amongst the most successful in this respect are those species that colonize the tops of walls, roofs and cracks in pavements. By virtue of being extremely parsimonious with the use of water, to the extent that nocturnal dew can sustain them through long droughts, they are tougher than their delicate appearance might suggest. The more resistant species in these desiccated habitats generally survived the prolonged drought remarkable successfully, but the less robust species from riverbanks and damp woodland still showed extensive damage, even after the wet winter of 1989/90.

The scale of these changes to mosses and liverworts was brought home to me even more forcibly in the summer of 1990, which once again brought with it a prolonged period of drought. I was commissioned to survey a flora of an area of land that would soon become the site of industrial development, in preparation for a public enquiry. The land had been surveyed once before, in the autumn of 1988, and when I repeated the survey I found it hard to believe that I was actually standing at the same location as the 1988 survey. Then, one of the main points of botanical interest had been the dense carpet of mosses and liverworts on the soil surface. Now, two years of drought had all but exterminated them.

One, or even two, summer droughts will not sound the death knell of our moss and liverwort flora, and if our climate were to become gradually drier, with barely perceptible temperature rises spread over a number of years, delicate plants might well have some time to adapt. Mosses, liverworts and ferns are all extremely mobile, because they are dispersed as microscopic spores that can drift for miles on air currents. They might slowly retreat from the drier southern areas and exposed habitats, but continue to thrive or, in the case of some species, even expand into cooler, wetter northern areas.

But consider what would happen if we had five summers like those of 1989 and 1990 in quick succession. Delicate plants would have no chance to recover, even in a wet winter, and would become locally extinct; only the toughest species would survive.

Some ferns would probably undergo the same traumas, because they have an unusual, two-phase life style. The familiar fern plant, with feathery fronds, is often an extremely drought-resistant organism. The rusty-back fern will thrive in the arid mortar of railway bridges and has often been referred to as a 'resurrection plant', capable of almost miraculous recovery from extreme desiccation after a brief shower of rain. The elegant little spleenwort ferns share the same kinds of habitats.

But these leathery fronds are only the most visible half of the life cycle. They release spores which germinate on wet soil to form a small, flat, fragile layer of green cells, less than half the size of the nail on your little finger. This tiny, cryptic part of the fern life cycle is where sex takes place; where the swimming spermatozoids fertilize the egg cells. Without permanent and intimate contact with wet soil it soon withers; dry springs might well eliminate this delicate stage of the life cycle from arid areas of the south east.

The behaviour of mosses, liverworts and ferns represents a sensitive litmus test for what lies in store for plant life in general. Few people have studied their physiology in sufficient detail to forecast their future, but there can be little doubt that the drier, south-eastern half of Britain will lose species, while the increasingly wet north-west will become their most important refuge.

Mosses and liverworts are so neglected by naturalists that most have never acquired common names and are only known by tongue-twisting Latin epithets. It will be well worth getting down on our hands and knees to get to know them better. When the moss in your lawn disappears and you are not using a weedkiller, it is time to take climate change seriously.

Extinction, precipitated by changes in temperature and rainfall, has always been an essential part of the evolutionary process. Cemented into a churchyard wall, a few miles away from my home, is the root system and part of the trunk of a tree which once formed a dominant element in the Earth's vegetation during the Carboniferous period, when the coal measures were laid down.

About three hundred million years ago, large areas of Britain were tropical swamp. It was a landscape of mud banks and shallow pools of water, inhabited by primitive amphibians. Towering overhead was a peculiar form of tree called *Lepidodendron*, quite unlike any alive today. Cones at the tips of its branches released clouds of spores that germinated on the mud banks to produce smaller structures with egg cells and swimming spermatozoids, much like the ferns of the present day. After a spermatozoid had fertilized an egg cell, a new *Lepidodendron* tree grew towards the forest canopy.

Within another fifty million years, *Lepidodendron* had disappeared, a prehistoric victim of climate change.

As the climate warmed at the end of the Carboniferous and throughout the succeeding Permian period, plants like *Lepidodendron* were doomed. Droughts became more frequent, so their spores were unable to germinate over large parts of their range. Without a surface film of water, spermatozoids perished. Eventually, as the swamps dried, the trees themselves slowly succumbed. Spectacular fossils, like the one in my local churchyard wall, are almost the only evidence of their existence.

Almost, but not quite. Club mosses, their direct descendants, still inhabit the mountains and moorlands of northern England and Scotland. While their giant antecedents in the tropical swamps suffered extinction, these smaller relatives somehow found suitable refuges that would sustain them for the next two hundred and fifty million years.

Many of our present-day club mosses, which in strict botanical terms are only very remotely related to true mosses, arrived here as the glaciers retreated at the end of the last Ice Age. As the ice sheets, some over a kilometre thick, slowly melted and retreated northwards, a characteristic alpine flora reinvaded the tundra that they left behind. Club mosses figured prominently, crossing the land bridge from Europe in the few thousand years between the melting of the last glaciers and the flooding of the English Channel.

As the climate in southern England improved, the club mosses and their associated flora moved north, to occupy the uplands and mountain tops. Cold, wet weather with short summers and severe winters suited them well. Water, so essential for their spermatozoids, was plentiful, and the severity of the climate kept at bay the less hardy plants that had followed them over the Channel land bridge before the waters closed and Britain became an island. While many of the colourful flowering plants needed insect pollinators to act as sexual go-betweens, carrying pollen from one flower to another, rainwater performed the same task for the spermatozoids of club mosses. At high latitudes and altitudes, where pollinating insects were less common, sex under water still gave club mosses a fighting chance in competition with their evolutionary successors.

For the last ten thousand years or so they have thrived in their upland refuges, but now, like their *Lepidodendron* forbears, they are once again threatened by a changing climate. Higher annual temperatures and long hot summers will push them into decline. Competitors from the south, better adapted to seasonal drought, are waiting in the wings to take their place.

Unless you are a dedicated botanist, it is a little difficult to get too worked up over club mosses. A charitable description of fir club moss, one of the rarer British species, is that it looks like a small, green spout cleaner which occasionally sprouts yellow, purse-shaped spore cases at the base of its nondescript, spiky leaves. Curiously enough, many people will be more intimately acquainted with the spores of stag's horn club moss, which are dusted on at least one brand of condom to stop the rolls of latex sticking together.

If the current phase of climate amelioration were to extinguish these last British survivors of a line of plant evolution that began long before dinosaurs roamed the Earth, few people would know or care.

But watching what happens to club mosses is important, because they represent a metaphor for our whole flora. Some species are a component of the so-called arctic-alpine plant community. This group of plants was in the vanguard of reinvading plant life at the end of the last Ice Age and includes some of our most beautiful wild flowers. Spring gentians, with small vase-shaped flowers of the deepest cerulean blue, that appear as the snows melt in Upper Teesdale. Moss campion in the Scottish mountains, with compact tussocks of leaves covered in deep pink flowers. Mountain avens in the limestone uplands of Yorkshire and Snowdonia, like snow-white buttercups springing from a carpet of miniature oak leaves.

As a schoolboy I spent much of my time searching for wild flowers, armed with a copy of *The Pocket Guide to Wild Flowers* by David McClintock and R.S.R Fitter. The appeal of this particular guide was that rarities were ranked with a star system. Virtually all the alpine species were two- or three-star rarities. I can remember struggling to the summit of Ben Lawers in a torrential rainstorm of the kind that only Scotland can deliver, to see some of these alpine gems. On the way up I passed one-star plants like yellow mountain saxifrage and two-star rarities like moss campion. Search as I might I never located any of the three-star rarities like alpine milk vetch and Highland fleabane. I can still remember the sense of disappointment as I made my way down again, cold and soaked to the skin.

The fact is that these alpine specialities have long been hard to find. Even with the rigours of a Highland climate they are at the southern boundary of their arctic range, on the edge of a climate that is almost too temperate for them.

Warmer summers and milder winters will do them no favours. They will retreat further north and ascend higher. Those on south-facing slopes may suffer direct damage from drought, but the real threat is competition from other species that will be favoured by higher temperatures and milder winters.

There is a north–south divide in the British flora which has developed with the reinvasion of the land mass after the retreat of the glaciers. To the north are the arctic-alpines and a range of other species that thrive in a colder, wetter climate.

Trees like bird cherry, for example, are rarely seen south of Yorkshire. Oyster plant, a rare, coastal species with fleshy, grey-green leaves and captivating tufts of deep blue flowers, is never found further south than Norfolk and is characteristically a plant of shingle banks on Scotland's coastline. As the climate warms, both species can be expected to retreat still further northwards.

Meanwhile, in the south, the late-comers to the flora, which scrambled across from the Continent before the Channel bridge flooded, need warmer summers for optimum growth. It seems more than likely that these will be the chief beneficiaries of a warmer Britain. Take the small-leaved lime, for example.

Studies of subfossil pollen deposits in peat bogs have revealed that seven thousand years ago, when the climate went through a wetter and warmer phase, small-leaved lime was a common tree throughout much of England. Today it is relatively scarce and does not regenerate naturally in the northern half of Britain. The reason for this is that over much of the country early July temperatures are too low for the tree's pollen to function properly, so it does not produce fertile seed and its range barely extends into Yorkshire. Warmer summers should lead to this species advancing into England's northern counties again. It set seed particularly well in the summers of 1989 and 1990.

Southern sun-loving species will expand to fill the niches left by the retreating northern plants, but aliens may fare even better. These species, from virtually every continent on Earth, have either arrived accidentally, through trade and travel, or have been deliberately introduced by horticulturists. Alien species outnumber native British species by a factor of ten to one. For many of these arrivals, which have perhaps survived here but not thrived, milder winters and hotter summers present unrivalled opportunity for expansion. There are plenty of precedents for what may happen when aliens make themselves at home.

Species introduced from hotter climates that now have a tenuous toehold in our countryside are bound to find milder winters to their liking and are expected to spread. Rising temperatures and changing rainfall patterns could push some native plants that are particularly adapted to a cool, moist climate into local or regional extinction. The arctic-alpine element in the flora and the mosses and liverworts are only two of several plant communities at risk. Vacant environmental niches are bound to become available for introduced species that will already have been given a boost by less severe winters.

The ability of alien plants to exploit available niches in our countryside is well illustrated by the cases of Himalayan balsam, Japanese knotweed and giant hogweed from the Caucasus, all of which have hopped over the garden fence and become rampant interlopers amongst the native flora. The last two have done so well that they were outlawed in the

1981 Wildlife and Countryside Act. They are now notifiable weeds, so that landowners not only must not plant them but also have a duty to control them.

There can be few lawns in the country that have not been colonized by the blue-flowered slender speedwell, another Caucasian native that was originally introduced as a rockery plant and escaped in the early part of this century. Curiously, this plant very rarely sets seed, only producing sporadic fruit capsules in the warmest of summers. It has achieved its status of gardeners' enemy amongst all those who treasure billiard table lawns purely through its astonishing capacity for vegetative spread. One pass of a lawn mower produces hundreds of cuttings, all of which have the potential to take root. Slender speedwell represents a cast-iron excuse for not bothering to mow the lawn.

While these and other alien plants have already blazed a trail through our countryside, other less hardy species from warmer climates look set to follow suit if milder winters aid their survival and hotter summers create a convenient niche.

The identity of the aliens that will inherit large areas of British countryside is a matter of conjecture. A few naturalists have taken a keen interest in these accidental or deliberate introductions. In 1919 Miss I.M. Hayward published a book entitled *The Adventive Flora of Tweedside*, a comprehensive catalogue of alien plant species that had reached the banks of the River Tweed via imported wool. Her book contained descriptions of weed species from as far afield as Australia and South America, all of which flourished briefly then expired as the wool industry waned.

Another botanist, Stephen Troyte Dunn, who worked in the Kew Herbarium, compiled a long list of records of garden plants that had escaped into the wild. An Imperial posting as Superintendent of the Botanical and Afforestaion Department in Hong Kong led him to rush into print with this list in 1905, hurriedly dictating the work to his wife as he packed for the journey to the Far East. Unlike Miss Hayward's aliens, Dunn's garden escapes have generally survived and many have flourished. Interestingly, he makes no mention of slender speedwell, which did not begin to assert itself as a garden escape until 1927.

These early works which catalogued escaped introductions were followed by numerous references in county floras to exotic plants which had become locally established; within the Botanical Society of the British Isles a specialist group of botanists still exists which records the appearance of any new interlopers that appear on rubbish tips, railway lines and around ports. But for the most part these studies have been made mainly for the pleasure of recording and identifying unusual plants. There has been very little detailed work on the dynamics of introduced species, attempting to measure their rate of increase or competitive ability compared with native species. The few studies of this type that do exist have been conducted only after introduced species have become well-established and troublesome weeds. There are some spectacular examples.

Rhododendron ponticum was introduced from cool, mountainous areas of Lebanon and Turkey in 1765 and was an instant success with owners of large country estates. The leathery-leaved shrub took to our cool, wet Atlantic climate and quickly spread, generating sheets of purple flowers which delighted the eye and offered good cover for game birds. But delight turned to consternation when the plants seeded themselves wherever disturbed soil offered the faintest prospect of a root hold.

They seeded themselves with a vengeance, with each flower yielding five thousand minute, wind-blown propagules. Once established, the dense canopy of dripping, dark olive-green leaves shaded out all other vegetation. *Rhododendron ponticum* was here to stay. No one knows for certain how much of it there is, but in 1988 it was estimated that it would cost thirty million pounds to clear it from the Snowdonia National Park alone. Within two centuries this thug of a shrub has progressed from a stately home status symbol to loathsome weed.

Aliens which have become serious weeds have certain characteristics in common. One is that they often have mobile seeds

SEX, ALPINES AND ALIENS

(*Rhododendron ponticum* and Himalayan balsam), another that they regenerate rapidly from fragments (Japanese knotweed and slender speedwell). Equally significantly, they have almost all, at one time or another, been favourite garden plants, ensuring their rapid and widespread distribution throughout the country. Gertrude Jekyll, the great Victorian garden designer, was a fan of giant hogweed, enthusiastically recommending it as a plant for the bog garden. She must take her share of the blame for its current status in the British Isles.

But some aliens owe their success, not to gardeners, but to modern modes of transport. Pineapple weed, whose pineapple-scented, petal-less flowers seem to be a feature of most farm gateways in Britain, is a Chinese plant that arrived here via the United States. Its seeds are thought to have been carried in mud on motorcar tyres. There is a popular (but perhaps apocryphal) belief that it owes its presence in gateways to vehicles reversing in to turn round. Its ability to thrive in the open, hard-baked habitat of farm track mud is likely to be a major contributory factor to its predilection for field and farmyard entrances.

Oxford ragwort owes its current distribution to the Great Western Railway. A native of volcanic soils, of the kind found on the slopes of Mount Etna, it was planted in Oxford Botanic Garden by Linnaeus, the great Swedish botanist who is responsible for our current method of plant taxonomy and who visited these shores in 1794. By the time the Great Western Railway's gleaming green locomotives began to serve the city, Oxford ragwort was established on local walls, poised to release its feathery, parachuted seeds into the slipstream of express trains that thundered along the tracks between Paddington and Penzance. The raked-out cinders of steam engine fireboxes suited the germinating seeds admirably, resembling the volcanic clinker of their Mediterranean homeland. By the time the age of steam had passed, Oxford ragwort had penetrated to the far corners of the railway network.

In the light of the epic odysseys of these introduced weeds, which of our current crop of garden plants is waiting to take advantage of a changing climate, vault over the garden wall and run amok in the fields and hedgerows? In this country there appears to have been little effort to predict the impact of a changing climate on garden plants that are known to have escaped into the wild and established a bridgehead. But in the United States the potential threat is being taken seriously.

Thomas Sasek and Boyd Strain of Duke University, North Carolina, have taken a close look at the climatic response of two introduced vines, Kudzu (*Pueraria lobata*) and Japanese honeysuckle (*Lonicera japonica*). The latter, a black-fruited version of our native honeysuckle, is established here, in Devon. Both species were introduced into the US from Japan in the early nineteenth century, as ornamentals, and were later cultivated to control soil erosion. Both are now serious weeds in the south-western US, eliminating the native flora wherever they occur. Sasek and Strain evaluated the indirect effects of atmospheric carbon dioxide enrichment, via climate amelioration, and the direct effects of greater atmospheric levels of carbon dioxide on these aggressive plants.

Current winter temperatures on the northern limit of these two vines in the US are too severe for them to make further progress up the continent, while droughts in the states to the west of their present stronghold are equally limiting. But given a physiological boost from higher levels of atmospheric carbon dioxide and a warmer climate, Sasek and Strain predict that both vines will again be on the march. Both make more efficient use of water as carbon dioxide levels rise, producing taller leafier plants, and while still drier climates at their western boundary might restrict any further inroads in that direction, they are likely to penetrate several hundred kilometres further north. From the boondocks of Louisiana, Japanese honeysuckle seems poised to march to Washington and to penetrate as far north as the Great Lakes. Perhaps our escaped Japanese honeysuckle in Devon will make similar territorial gains as the climate changes.

SIX

Frosts Are Slain

> Frosts are slain and flowers begotten,
> And in the green underwood, blossom by blossom,
> The Spring begins.
>
> SWINBURNE

If you fancy yourself as a 'countryman', there is one talent which is absolutely indispensable. This is the black art of convincing town dwellers that years of rural living have left you with an innate ability to forecast the weather from natural signs.

There is a certain satisfaction in leaning over a farm gate, sniffing the air, running a shrewd eye over a hedgerow heavily laden with berries, ruminating for a moment or two and then predicting with a knowing nod of the head that 'We are in for a hard winter'. I suppose this is a lingering echo of former civilizations, where soothsayers and oracles were a respected part of everyday life. We all know that it must be a load of bunk, but it has become a kind of ritual, like reading horoscopes in the newspapers.

There is a relationship between good berry crops and weather, but it has more to do with the past than with the future. A good holly berry crop, for example, depends to a large extent on the weather in May and early June, when honey bees and flies pollinate the tiny, waxy-white flowers. There are separate male and female holly bushes, and without insects to carry the pollen from one sex to the other there would be no berries. Cold, damp weather in late spring and early summer, preventing bees from foraging, will leave bushes without berries for Christmas.

SPRING FEVER

Changes in climate have the potential to reset the seasonal clock, disrupting the pattern of plant growth and reproduction. If the recent pattern of short, mild winters persists, the onset of spring may begin to look quite different.

The diurnal change in temperatures during a typical English summer's day might be of the order of 20°C, so an increase in mean temperatures of one or two degrees seems insignificant. Most plant growth occurs between 10°C and 35°C. Within this range higher temperatures lead to faster growth. Even a small mean change, of perhaps one degree, can have a dramatic effect on rate of growth and a very significant influence on flowering and the production of seeds.

There is a general expectation that it is in trees that we will first notice a difference in the spring countryside. While short-lived plant and animal species may be able to evolve at a rate that will allow them to cope with climatic changes as they occur, a tree with a life span of a century or more may go into an almost imperceptible decline as the climate becomes less favourable. As an individual it may grow or even thrive in warmer summers, but a change in the pattern of climate is likely to affect the vital processes of flowering and seed production.

Amongst those early-flowering trees that will take advantage of milder winter conditions to produce precocious flowers, poor pollination conditions could have a disastrous effect on seed set. Late frosts can sterilize pollen in premature flowers. Hazel catkins at Christmas might mean no hazel nuts in November and no food for the animals that depend on them.

In the spring of 1990 I carried out some revealing experiments on the dangers of precocious flowering for hazels. A hazel catkin begins life in late summer, as a reproductive bud. This develops into a cluster of miniature green, perfectly formed catkins by the time the hazel leaves change from green to buttery yellow in October. Then they sit through the winter, ready to respond to warm spring days by lengthening, turning yellow and shedding pollen.

I took some catkins that had been coaxed into maturity by

the mild winter and early spring of 1990 and treated them to the same temperatures that they would experience in a night of moderately severe frost. The effect on the pollen fertility was dramatic. Over three quarters of the pollen grains were killed by this single cycle of freezing and thawing. A few weeks after I carried out these tests there were natural severe frosts in April, which devastated fruit tree blossom in many parts of the country. It would be interesting to know what effect it had on hazel nut crops in hedgerows.

Plants like hazel take a calculated evolutionary risk. Hazel pollen is carried in the wind, to tiny purple stigmas that sprout from buds. Hundreds of thousands of pollen grains are needed to secure a single pollination, because the pollen is diluted in the wind as it leaves the catkin. The presence of leaves on the twigs would lower the likelihood of a successful pollination still further, so hazel sheds pollen before the leaf buds open, before the female stigmas are masked by foliage. Nut production depends on the delicate balance between the need to secure successful pollen transfer and the risk of frost damage.

Early flowering shrubs often suffer from the effects of sudden drops in temperature. Blackthorn, which produces a mass of white blossom in spring before its leaves unfurl, is often scorched by frost. By flowering early it monopolizes bee pollinators as they emerge from hibernation, but unseasonably cold weather can discourage these from foraging. So good crops of sloes depend on a fortuitous balance between frost-free nights when blackthorn blooms and warm sunny days, which encourage bee activities.

Mild winters and early springs are already recognized as a cause of apple crop failure, by promoting premature flower development. Early blossom in fruit trees is particularly vulnerable to late frosts. But even if these do not materialize, mild winters can move flowering out of step with availability of bee pollinators; by the time that sufficient numbers of bees are ready to pollinate, the peak of flowering will already have passed. Studies carried out on Cox's Orange Pippins in 1980 suggested that a one degree centigrade increase in mean maximum temperature in the spring period would decrease apple yields by about one tonne per hectare. While the apple crops in the south would suffer, it might not be such bad news in the north of England, where less severe springs might make planting an apple orchard a worthwhile proposition at last.

For some trees, a mild winter will not induce early spring growth. A significant proportion of tree species, including beech and sitka spruce, need to experience a minimum number of days at low temperatures before their buds will emerge from dormancy.

Trees form dormant resting buds in winter for several reasons. Leaf shedding in deciduous species, after first withdrawing useful nutrients back into the branches and trunk, allows the tree to discard foliage that might be directly damaged by winter conditions. In the low temperatures and light levels of winter the leaves would not carry out their essential function of photosynthesis, converting atmospheric carbon dioxide into sugars, with any great efficiency so there would be little net benefit in the tree retaining them. More importantly, the foliage would continue to evaporate water into the air, so the tree might actually suffer from acute drought in winter when soil moisture was frozen. So it is usually in the interests of a tree to discard its leaves when the short days arrive in autumn and to wrap next year's embryonic leaves in tightly overlapping bud scales, that shield them from frost and drying winds.

Evergreen conifers do not shed their needles synchronously, but drop them in a continuous rain. Their foliage has a number of adaptations which allow the green tissue to carry out photosynthesis even at low temperatures and which also minimize water loss through the leaf surface. So it pays conifers to retain their foliage throughout winter; it is no mere coincidence that they are the dominant trees in boreal regions. There, they are supremely adapted to life at low temperatures. But like deciduous trees, they must protect their delicate bud tissues against the worst that the climate can deliver and they too form dormant buds, which are only triggered into growth when the weather conditions improve.

If this release from bud dormancy merely depended on a rise in temperature or daylength, the outcome could be fatal. A late frost could kill all the opening buds. Instead, buds of different species contain varying amounts of chemical compounds which inhibit growth. The compounds are slowly destroyed by low temperatures and trees have a minimum requirement for successive days of chilling before the dormancy compounds are broken down and the buds can burst into life.

Paradoxically, mild winters will delay bud burst of beech and sitka spruce and when they finally struggle into leaf some of the growing season will already be lost. The significance of this, in terms of lower annual growth rates and economic loss, is hard to predict because warm summers may produce some compensatory growth, provided that sufficient soil moisture is available.

The forestry industry is well aware that this is a major potential problem posed by changing climate. Even now some introduced coniferous timber trees, like sitka spruce, receive close-to-minimum essential levels of low temperature chill for normal bud burst in some parts of the country. A few extra warm winter days and their spring growth will suffer.

Tree planting is fashionable. In 1989 six hundred thousand trees were planted in National Tree Week. Farm woodland schemes, as an alternative land use for fields taken out of cultivation, are being actively encouraged. Several new Community Forests are planned. The climate that these trees will experience when they approach maturity in the second half of the next century will be very different from today's. Foresters are already beginning to recognize that the seeds sown now, which will produce forests for the twenty-first century, should be taken from trees growing in the same type of climate that their offspring will eventually encounter as they mature. If, as predicted, the south coast of England will have a climate approximating to that of southwestern France, we need to plant saplings raised from seeds harvested in France and not from the New Forest.

This has an interesting knock-on effect on conservation policy. It is now conventional wisdom amongst conservationists that we should plant seeds, in amenity tree planting schemes for example, that are of local provenance wherever possible. The arguments for this are largely academic and based on the need to conserve the genetic structure of locally adapted populations. This could be a dangerous principle to adhere to if we take into account long-term climatic change. Trees grown from seed of local provenance could be genetically obsolete in the climate of the twenty-first century.

Our current native flora reinvaded Britain at a time when glaciers were retreating. Between seven and ten thousand years ago large areas of the landscape would have resembled tundra, with fast flowing shallow rivers, short summers and long, savage winters. So it comes as no surprise to find that much of our flora is adapted to a strongly seasonal climate where long periods of frost and snow are regular features of winter.

Ice presents plants with serious problems. If ice crystals form in cells, the delicate cell membranes are killed. When dahlias are reduced to a liquid, pulpy mass by a heavy frost, it is because jagged ice crystals have burst their cells; the liquid cell contents leak out and the plant collapses like a deflated balloon.

Paradoxically, freezing conditions also cause drought. As far as a plant root is concerned, soil full of frozen water is about as useful as soil with no water at all.

So the best strategy is usually to shut up shop for the winter. As the days shorten perennials withdraw all the useful nutrients from their leaves into there stems and trunks, producing a temporary, spectacular autumn blaze of colour from the residual pigments in the leaves before they are finally shed.

Annuals employ an alternative strategy and die, leaving a bank of dormant seeds in the soil to sit out the winter and produce a new generation in spring. The key to success here is to make sure that the seeds germinate at the right time. Too early and they will be killed by ice and snow, too late and valuable growing time will be lost.

The biochemical mechanism which acts as a clock in seeds seems to be similar to that which prevents buds from bursting too soon. Many of the details of its workings are still unknown but there are certainly chemical inhibitors produced in seeds which prevent them germinating in the depths of winter. These are usually produced shortly before the seeds are shed.

Anyone who has attempted to raise wild cowslips will know that the seeds need a winter's cold treatment before they will germinate. What is less well known is that they will germinate immediately in late summer, if they are harvested and sown about a week before the seed capsules mature. The dormancy clock is set in the last few days before the capsule splits. There is a strong suspicion that the dormancy compounds are sequestered in seed coats; if you peel these from sycamore seeds they will certainly germinate faster, without their normal requirement for a cold chill.

The seeds of at least a quarter of the British flora need a cold chill before they will germinate. What will happen to them if milder winters produce a frost-free southern half of England? Would the seeds be doomed, like the Sleeping Beauty, to sleep forever, until they were woken up by the icy kiss of winter?

A probable scenario would be that there would be a decline in the species with these dormancy mechanisms, as a result of germination failures. But it most instances it would be temporary.

Most populations of living organisms are extremely variable, not only in their visible characters but also in most of their hidden ones. They have a cryptic reservoir of genetic variability which allows them to evolve as conditions change. Genes which are useless or perhaps even positively harmful today may be essential equipment for the climate of the twenty-first century.

Similar hidden variability is carried in human populations;

in extreme instances it expresses itself in congenital birth defects, while more subtle physiological variability occurs in characters ranging from sensitivity to antibiotics to the ability to smell the scent of Freesias. Similar defects and variations occur regularly in plants, but they usually pass unnoticed unless they are deliberately looked for.

The cold requirement for seed germination is one such variable character. Plant breeders often remove it altogether by selecting rare varieties which do not make the seed dormancy chemical. Some wild carrot seed will only germinate in spring, after exposure to a winter of low temperatures: cultivated carrot seed will germinate uniformly at any time of the year, provided that they have adequate moisture and warmth.

One example of this variation in cold requirement occurs in Himalayan balsam, an alien annual which originated besides fast flowing streams in the Himalayan foothills but which is now common beside canals and rivers throughout Britain. The plants are very sensitive to freezing temperatures and are killed by the first frosts, while the cold-hardy seeds normally have prolonged dormancy requirements which prevent them germinating in the depths of winter.

Pauline Mumford, at Birmingham University, has studied this introduced species in detail and has discovered that it needs at least six weeks of temperatures below 4°C to overcome its innate seed dormancy. It also has a fail-safe mechanism, whereby a warm period in the middle of a cold spell will reset the dormancy clock back to zero. This prevents seeds germinating precociously in mild periods during winter, a fatal mistake that would inevitably lead to the death of the seedlings in subsequent frost. The plant also has a second safeguard, in that the seeds remain viable for at least three years, even when they have absorbed water, so if they do not get a sufficient cold stimulus in the first winter they get a second chance in the next.

But the overall dormancy requirement of a given population of seeds is also variable. Some seeds germinate after forty-five days at 4°C, others do not stir until they have experienced eighty-nine days of low temperatures. So, with a combination of a mechanism to prevent precocious germination, a mechanism to accumulate a dormancy-breaking cold stimulus in instalments spread over two or more winters and a pool of genetic variability that can alter the length of the cold period needed, Himalayan balsam seems to have catered for most of the possible permutations of winter weather.

The germination requirements of Himalayan balsam have been well studied, but much less is known about other cold-requiring seeds. It seems likely that many may have simpler control mechanisms. This would mean that many seeds would either fail to germinate, through too short a cold period, or might germinate too soon in mild weather and die in a following frost. But amongst the pool of genetic variants within any given species there should almost always be a few seeds with the right genetic characteristics to ensure their survival and that of the species.

Within all tree species genetic variation for the ability to respond to earlier springs certainly exists. A quick survey along any length of hawthorn hedge in late March will reveal individual plants that are already in full leaf, while a similar survey in late April will reveal others that have yet to loosen their bud scales. The same pattern of variation is also apparent for flowering time, with some hawthorns blooming much earlier than others.

The most extreme and most famous variant in this respect is the Glastonbury thorn, a mutant hawthorn which flowers in January and is reputed to have sprouted from Joseph of Aramathea's staff on the site of Glastonbury Abbey. This is a precocious flowering mutant, comparable to the variant of the cultivated cherry *Prunus subhirtella 'Autumnalis'*, which is almost always in full bloom for Christmas.

So plants with seeds or buds which need prolonged low winter temperatures to coax them back into growth may undergo a temporary decline, until rare gene mutants with different physiological requirements have had time to multiply and spread. Their success would depend on whether they could reassert themselves amongst the vegetation with less exacting requirements that will have taken over while they were temporarily readjusting their clocks.

The other key factor which may determine which plants will be able to exploit their reservoir of genetic variability and evolve to keep pace with changing winter weather or, indeed, changing climatic conditions at any other time of the year, is length of life cycle. A groundsel plant can easily get through six or more generations per year, producing thousands of seeds after each flowering. An English oak may not produce its first acorn until it is forty years old. The unprecedented rate of climate change which we are about to witness will mean that rapidly reproducing annual weed species will be at a very distinct advantage in the evolutionary race to adapt to shifting weather patterns.

There is plenty of evidence that weeds can adapt rapidly to changing environmental conditions. Scientists at the University of California have recently studied the rapid evolutionary potential of annual meadow grass, a ubiquitous weed which everyone must have in their garden here in Britain. They chose their local golf course at Davis as their laboratory and showed that this tenacious little weed can reprogramme its whole life history strategy in response to temperature and supplies of water.

In the roughs the grass behaved as an annual, growing fast and setting seed before Californian summer droughts struck. On the greens, where the grass is mown to a regulation six millimetres and supplied with luxury levels of water, it evolved into a perennial, putting its energies into leaf production rather than seeds. The switch from annual to perennial happened within the twenty-five-year history of the golf course.

Evolutionary change of this kind will certainly produce a response to shifting climate patterns. The end result will be different kinds of vegetation from those that we are now familiar with. As summers become warmer, we may be about to enter a century where weedy species inherit the Earth vacated by the slow growing, ecologically demanding species which can not move with the times. Rapid climate change may well provoke an ecological convulsion in the countryside.

SEVEN

The Tangled Bank

> It is interesting to contemplate an entangled bank, clothed with many plants of many kinds, with birds singing on the bushes, with various insects flitting about, and with worms crawling through the damp earth, and to reflect that these elaborately constructed forms, so different from each other and dependent on each other in so complex a manner, have all been produced by the laws acting around us.
>
> CHARLES DARWIN, *The Origin of Species*

Plants, like people, are very competitive organisms. They may not seem so, because for the most part their daily movements are confined to growth, a slow process which only dedicated plant watchers can really appreciate. But within the confines of the communities in which they exist, plants are engaged in a continuous struggle to optimize their performance. In this, they are constrained by the climate, the soil, the animals which feed on them, the pathogens which infect them and the other plant species which they coexist with.

This combination of factors eventually leads to a stable ecosystem, with all of its component organisms balanced in a precarious equilibrium. Should any of these intermeshing factors change, then the stability of an ecosystem is affected and changes in species composition take place. New species dominate and those that were once common may decline or even disappear.

There are at least two schools of thought about the origin of the different types of vegetation which we recognize in our countryside. One camp believes that the component species of ecosystems, like chalk grassland or hay meadows, have co-evolved within the set of physical, climatic conditions that they now experience, making a sort of evolutionary deal where they divide up the available resources amongst themselves while making the minimum of overlapping demands. This concept portrays an ecosystem almost as a living organic entity, with the individual components acting like parts of a body.

It is an idea that has a certain appeal, especially for those who like to classify vegetation. It implies that each vegetational 'organism' has a long history and is likely to continue more or less unchanged, provided that the physical constraints, like temperature and rainfall, remain more or less constant. So, today's limestone grassland species composition will be tomorrow's also; or would be, were it not for the fact that the climate is changing.

This concept forms a basis for the classification of different vegetation types which is routinely used by nature conservationists when they decide which types of vegetation need to be managed or preserved.

The opposing camp takes a more macho view of events. They say that each species in an ecosystem is distributed within the general vegetation entirely on the basis of its own physiological requirements and is only prevented from becoming dominant by its competitive interactions with other species in the vegetational system that it finds itself in. It is a variant on the 'nature-red-in-tooth-and-claw' outlook on life, adapted for the world of plants, and it would predict that vegetational composition is fluid and has always been so. Each type of plant ecosystem recognizable today is really in the process of dynamic change and the fact that we can classify them now on the basis of the particular species that constitute them has little long-term relevance. Today's classifications are really snapshots in time, a still frame in an endless video of constantly changing ecosystems.

In reality, as with many scientific arguments, there may be elements of truth in both hypotheses. The fact that controversy exists reflects a limited state of knowledge of the forces that maintain particular forms of plant ecosystem.

Recent evidence from studies of subfossil deposits of pollen, when researchers have mapped the abundance of different kinds of pollen over long timescales during which climate has changed, reveal that the second hypothesis may be the more accurate. The dominance of different species of tree in several North American forests has clearly changed over the last ten thousand years and today's natural forests have a quite different species composition from their primaeval forbears. Species, it seems, wander in and out of different types of vegetation, searching for conditions that suit their particular physiological needs.

This concept of a peripatetic plant species is hard to swallow unless you take into account the gradual pace of most natural climate change and the ability of plant species to disperse themselves.

I remember once walking through fields after a recent heavy snowfall and pausing about four hundred yards downwind of a line of sycamore trees. A stiff breeze was blowing and a constant rain of spinning seeds flew through the air and buried themselves in the soft snow. The dispersal ability of the sycamore was clearly measured there in the snowdrifts, like the marks left from a field athlete's discus. The trees were on the march.

Some trees with smaller, lighter seeds can disperse over much longer distances, then germinate, grow and set seeds again in the space of a few years. Birch is particularly adept at this and is so invasive that it is regarded as a weed by many foresters. As climate ameliorates there is every chance that its dispersal ability will mean that it will come to dominate habitats like heathland or even heather moorland at lower altitudes, unless steps are taken to control it. Birch, and perhaps some other trees, will begin retaking some of the high ground that they lost after the climatic optimum some five thousand years ago.

Even the largest plants with heavy seeds are remarkably

mobile and can enter and leave different kinds of vegetation surprisingly quickly, although few plant species will be able to move unaided between widely spaced geographical locations during the rapid phase of climate change which is now expected.

One interesting insight into the rate at which trees can migrate in response to temperature change comes from some research done in Scotland by Annabel Gear and Brian Huntley, two of my colleagues at Durham University. They located fossil stumps of Scots pine trees in the far north of Scotland and were able to demonstrate that pine forests covered the whole region for a brief period of four hundred years, about four thousand years ago. They believe that these forests were able to expand northwards by some eighty kilometres during a warm period when boggy land dried out, and then retreated again as the climate became colder and wetter.

Their data show that the Scots pines migrated at a maximum rate of about half a mile per year. Scots pine is quite a mobile species, with wind-blown seeds, and if this represents the maximum speed at which such trees can move in response to climate change, then forests will struggle to keep pace with a shifting climate. Gear and Huntley have pointed out that if the rates of tree movement that they have measured from past climate changes represent the limit of a tree's ability to move in response to rising temperatures, then this will be ten times too slow to keep pace with the current round of rapid changes.

This fluid model of plant communities would suggest that current climate change will produce striking modifications in the plant species composition of some of our most treasured wildlife habitats. Just what these changes will be is difficult to predict. One way to find out will be to watch plants very closely.

Lichens are perhaps the most nondescript and unspectacular members of our flora. While the study of mosses might be classed as a minority interest amongst botanists, the study of lichens is an even more select and esoteric occupation.

For botanists who tire of the flamboyance of flowers, lichens represent a refreshing digression. Map lichen, *Rhizocarpon geographicum*, encrusts rocks with irregular coloured patches that coalesce with lichens of other colours, like maps of long-lost empires. *Stereocaulon*, which grows on rocks in the exposed mountainous areas, looks like a collection of angry, bunched fists. Blood-drop lichen, *Haematoma*, is another rock-encrusting species that is covered with crimson blobs, like spilled blood at the scene of a battle. Pixie-cup lichen, *Cladonia pixidata*, covers bare soil with stalked cups, like golf tees, and *Graphis elegans*, which grows on bark, resembles random graffiti made with a soft lead pencil.

The slow growth of lichens has been used before as an index of environmental change. Circular colonies increase in diameter by a steady, slow increment and you can calculate their growth rate by measuring their diameter on grave stones. The date of death on a tombstone sets an upper limit on the age of the lichen colony, and from that it is easy to calculate average growth rates of different encrusting species. Once you have calibrated your lichen species in this way, it is a simple matter to measure the age of other objects that form substrata for their growth.

One application of this simple technology has been to measure the rate of retreat of melting glaciers as climates become warmer. Lichens also have an important ecological role in the formation of soils from rock. As a glacier melts and shrinks it releases rocks which the ice has swept up and held in its grip for thousands of years. Lichens are the first colonizers of these freed boulders, secreting acids which help to break them down and producing humus as they die. The size of lichen colonies increases as you move away from the edge of a glacier, so if you know how fast the lichens grow you can calculate how fast the glacier is melting.

Part of the fascination with lichens stems from their extreme resilience to harsh environments. These pastel-coloured encrustations on rocks can bake in blistering sunlight all summer long and then freeze under a layer of ice throughout the winter, without apparent harm. They grow at a snail's pace, especially those in exposed mountains and moorlands, which may only achieve a paltry millimetre of growth per year. But at least one dedicated lichen watcher of my acquaintance believes that he has detected a spurt of growth in some of these species during recent mild winters. He blames climate change.

I have a hunch that he might be right and that his patient observations might well be detecting the early symptoms of climate change.

There are two ways in which the changing climate might favour faster growth in these tenacious plants. One is simply that higher temperatures will stimulate physiological activity by speeding up the plants' basic biochemistry. The chemical reactions inside a cell work faster as the surroundings become warmer. Just like internal combustion engines, they have an optimum working temperature at which they are most efficient. Many lichens which are confined to sites of summer drought, like roofs and bare rocks, grow during the warmer days in winter, when water shortage is not a problem. So a warm, wet winter should boost growth rates.

The second factor which is likely to favour faster lichen growth is that they may benefit directly from the higher carbon dioxide levels in the air. Carbon dioxide gas is the raw material that plants use to manufacture sugars, the building blocks that are essential for growth. When carbon dioxide levels rise, photosynthesis, the biochemical manufacturing process which produces the sugars, usually works more efficiently.

A lichen is a partnership. Inside the grey-green, scaly thallus that coats branches in woodlands, or the brilliant orange or yellow circles that cling to tiles and slates, lies a mass of algal cells, trapped in a web of fungal threads. The algae that form one half of the partnership can also be found growing alone, usually in wet places, but the fungi depend on the intimate relationship for their survival. About a quarter of all fungi – some eighteen thousand species – are thought to be locked into an inseparable pact with their algal partners. You can separate a lichen into its fungal and algal components in the laboratory, but you cannot reconstitute it. The intimate relationship is cemented by the rigours of the environment.

The benefits of the association are clear enough for the fungi. Their presence makes the trapped algal cells leaky, so the sugars and nitrogenous compounds that they synthesize, using their green photosynthetic pigments and the energy from sunlight, trickle out and nourish the fungus. For the algae the benefits are less clear, although the relationship does allow them to exist in extreme environments where they might not otherwise pursue an independent existence. A combination of higher carbon dioxide levels and higher winter temperatures will enhance the growth potential of the algal component of this partnership and speed the growth of the whole organism.

My lichen-watching friend plans to use lichens to monitor climate amelioration. Twice a year, in spring and autumn, he measures the diameter of the circular lichen patches that he has designated as calibrators of climate amelioration. He is a man of infinite patience, which is fortunate, because throughout the 1980s some of his circular lichens only increased in diameter by nine millimetres. But now they are quite definitely growing faster, sometimes by up to three millimetres per year. For a lichen, this is tantamount to sprinting.

Professional botanists will undoubtedly use similar techniques to monitor changes on growth rate in flowering plants in natural ecosystems. By marking out fixed plots and measuring characteristics like flowering performance and leaf and seed production, it should be possible to plot changes in growth rate and the relative competitive ability amongst species.

Studies of the effect of extreme climatic events in single years can give some inkling as to the effect of high temperatures and prolonged droughts on individual plant species within plant communities.

When the now-famous hot summer of 1976 arrived, Brian Hopkins from New England College at Arundel in Sussex was engaged in a study of the effects of people at picnic sites on the plant species of chalkland turf at Goodwood Country Park. He was ideally placed to monitor the effects of long weeks of drought on the parched grassland.

The vegetation of chalk downland is a botanical delight. While still a schoolboy I spent a long summer searching the South Downs for orchids and became very familiar with the richness of the plant life in the short, sheep-grazed turf. The term 'living tapestry' has been applied to chalk grassland, and although it is a well-worn cliché there is probably no better way to describe it. Rabbit grazing generates dwarf races of plants like harebell, wild carrot, yarrow, salad burnet, clover, and wild thyme that are tightly woven into a dense carpet, with the spaces filled in by mosses. Constant chewing by grazing animals suppresses some of the most competitive species, so that plants which would otherwise be swamped by them can maintain a foothold. The end result is a remarkably rich flora.

In some places, where the turf is temporarily protected from grazing, dense swarms of bee orchids appear, or clusters of pyramidal and fragrant orchids. I spent day after day walking over this terrain in search of orchids and often sat on the steep sloping hillsides amongst this vegetation, eating lunch with the landscape stretching out below. Halnaker windmill in the middle distance, Chichester cathedral sitting on the coastal plain and the Solent beyond, glinting in the afternoon haze.

I digress. Hopkins' study followed the fate of this vegetation, which is just about perfect for picnickers, throughout the rainless summer of 1976. At Goodwood, sunshine levels were 140 percent above average, and the soil moisture deficit, a measure of drought, was at an all-time record. Throughout the summer the turf gradually withered, and by late August there was eight times more bare ground than in a normal year. Almost half of the ground in the study area became devoid of vegetation.

Sheep's fescue, the fine grass that contributes significantly to the unique character of downland turf, went into steep decline, along with the mosses, suckling clover, mouse-ear hawkweed and ribwort plantain. But some species were almost immune to the drought. Cock's foot grass, dandelions and rough hawkbit survived virtually unscathed.

Autumn brought a downpour. Torrential rain in September restored soil moisture levels so thoroughly that 1976 turned out to be a year of almost average rainfall, despite the months of drought. In fact, overall weather conditions were something like those that are predicted for the south-east in the next century.

As the water returned, the plant life began to recover. The sheep's fescue, mosses and ribwort plantain recovered well, but the suckling clover and mouse-ear hawkweed lagged behind badly. On the other hand, the species which had weathered the drought without decreasing began to increase. The coarse cock's foot grass and deep-rooted dandelions and rough hawkbit, together with daisies, increased their share of the available space. Generally, seedlings of the species in the daisy family did conspicuously well.

This short-term study of the vegetation of the thin soils that overlay chalk showed the dramatic effect of severe summer drought and the first stages in a change of species composition, with the rough cock's foot grass, hawkbits and dandelions profiting from the demise of less tolerant species. The drought was the culmination of sixteen months of unusually dry weather, but although species frequency oscillated wildly, no plants disappeared entirely from the study area.

While short-term studies of extreme climate, of the kind that Hopkins performed, are available for some sites, detailed long-term surveys of changes in vegetational composition are exceedingly rare.

But in a few cases such data already exists. Professor Arthur Willis has surveyed the changing floral composition of road verges near Bibury in Gloucestershire for over thirty years, as part of a long-term study on the effects of herbicides. His unsprayed control plots offer one of the best series of data on the effect of annual climatic variations on plant performance. Results show that some species, like smooth stalked meadow grass, respond very quickly to years of extreme climate; other species have increased gradually over the whole survey period. But untangling the direct effects of climate from the indirect effects of changing levels of competition between species and subtle, underlying changes in soil characteristics is a long, tedious and formidably complex statistical process.

The problem with studies of this kind is that, although they can reveal changes as they unfold, they take an inordinate amount of time and patience and are only of limited help in predicting the future. Even in thirty years, climate change has been insufficient to generate any obvious trends. If the measurements are carried on for another century they may provide an invaluable record of a sustained change but with the rapidity of impending climatic change, a more direct approach to the problem is required. We need answers now.

One method is to grow wild species in the laboratory, under controlled climatic conditions, and to measure their performance under the range of conditions that are predicted for our countryside during the twenty-first century.

Dr Phil Grime at Sheffield University's Unit of Comparative Plant Ecology has been doing just this for several years, as part of an intense screening programme. Some of his results have already revealed that several of our best-loved species will fare badly in a milder Britain.

Grime and his colleagues recognize two broad categories of plants. There are those which grow rapidly early in the year, like bluebells and celandines, and produce the blaze of woodland colour which is so characteristic of an English spring. They flower early because their cells have divided in the previous year and have lain dormant through the winter. they are ready to respond to milder winters by expanding rapidly in spring, pushing new shoots through the soil soon after Christmas.

In contrast, there are plants which carry out all their cell division in the same calendar year that they flower. They cannot perform the energetically expensive cell division process until temperature rise much further, so they form the summer flowering component of the flora.

One of the most alarming predictions to emerge from Grime's data is that milder winters may well close the gap in growing period between these two species, allowing those that need high temperatures for cell division to grow and flower earlier in the year. This may well mean that they will compete strongly with the spring flowers. The carpets of bluebells that are one of the glories of deciduous woodlands may become threatened by invading, competitive grasses and weedy species whose cell division and growth period has been advanced by a warmer climate. Then, if summer droughts were to curtail the cell divisions that bluebells must make in preparation for the following spring's burst of cell elongation and growth, the interlopers might take over.

Mild winters have produced some uncharacteristic growth phenomena in many plant species. Butterbur usually flowers before its leaves emerge from the ground, but after the mild winter of 1988/89 the leaves appeared at the same time as the blooms. Black knapweed leaves stayed green throughout the same winter. Whether these observations are also part of a long-term trend remains to be seen.

Some observers believe that climate-induced changes in vegetation composition will depend on much more than the direct effects of the physical environment on the plants themselves. They say that what happens below the soil surface is crucially important in determining what we see on the surface.

At the time, the droughts of 1959, 1975, 1976, 1983, 1984, 1989 and 1990 all seemed to be one-offs. But now we are faced with the prospect of a changing climate where such hot summers, coupled with wet winters, could become the established pattern. Predicting the effects that these new temperature and rainfall scenarios will have on the composition of plant communities involves much more than a simple assessment of the drought tolerance of the individual species concerned. The microorganisms that live in the soil could also play a decisive part in the eventual outcome.

Soil is far from being an inert medium for plant growth. In many ways it behaves like any living organism. Amongst the soil particles around plant roots there are billions of bacteria and interlaced wefts of fungal hyphae that exist in an intimate and poorly understood association with the above ground vegetation.

A significant proportion – perhaps 50 percent in some instances – of the carbohydrates and nitrogenous compounds that are made by the green leaves of plants during photosynthesis leak out of roots and sustain a vigorous association with the soil microflora.

In some cases, in legumes like vetches and clovers, the below-ground partners are bacteria, which convert atmospheric nitrogen into nitrate fertilizers, which sustain the plants. Sometimes the plant/microorganism partnership is less specific but equally valuable. Many below-ground bacteria protect plants from pathogenic organisms, while others break down soil and release minerals in a form that plants can absorb.

The most significant association between plants and soil organisms involves fungi, which either invade the outer cells of plant roots or form a sheath of fungal tissue around them, greatly enhancing a root's ability to absorb minerals from the soil and assisting in water uptake. Such fungi are known as mycorrhizae (literally 'fungal roots') and as many as 80 percent of all green plants form this type of association. Several thousand different fungal species are known to be involved. This plant-fungus partnership is particularly important in forest ecosystems.

Dr Dave Perry and his colleagues at the Department of Forest Science at Oregon State University in the US have focussed attention on what might happen to these mycorrhizal associations during a rapid period of climate change. They believe that the ability of plant species to move into and out of ecosystems may depend critically on the fungal and bacterial flora of the soil.

They argue that when ecosystems change, the soil flora changes too, making it unsuitable for plants that depend on specific mycorrhizal associations. In extreme cases, the soil can die, or become invaded with fungal species which actually inhibit plant growth. When this happens, opportunistic weedy plant species take over the bare soil, degrading the ecosystem in a way which is extremely difficult to reverse.

Evidence that this can happen comes from areas in southwest Oregon where forests have been felled, native vegetation has been sprayed with herbicides and the ground left for colonization by invading weeds. In these areas it has proved virtually impossible to reintroduce trees, because the soil fungi which were essential to their success have disappeared, and the habitat has been converted to poor quality annual grassland. Besides acting as intermediaries in the mineral nutrition of the trees, these subterranean microorganisms also maintained a good forest soil structure which rapidly collapsed once the fungi and bacteria were deprived of tree root hosts and disappeared. Now, only trees which are planted in introduced pockets of healthy, mycorrhizal forest soil will establish in the degraded environment.

Perry's portrayal of plant communities below ground level is very much along the lines of the community-of-cooperating-organisms school of thought, with the soil microflora regulating the competitive balance between species, and he points to convincing evidence that competition between plants is reduced by the presence of mycorrhizae. His experiments have shown that plants which would normally inhibit one another's growth can coexist and thrive if soil mycorrhizal fungi invade the roots of both, possibly because the fungi equalize the distribution of soil resources between the species. Mycorrhizae may play a key role in maintaining the health and stability of complex plant communities and protecting them from invasion.

If these subtle interactions between the soil microflora and plants are as widespread as Perry and others believe, the implications for the rate and direction of change in ecosystems may be profound. Loss of individual species could change the competitive balance below ground. If the change in fungal and bacterial flora proceeds far enough and fast enough, the delicately balanced community of plants will be invaded by aggressive weeds which do not depend on mycorrhizal interactions.

There is a further disturbing twist in this complex tale. Perry points out that the interaction between above-ground plants and below-ground fungi should help to stabilize plant communities during the rapid climate change. It is not uncommon for drought-sensitive species to be able to coexist alongside drought-tolerant species, for example, because they share fungal mycorrhizae which allow both species to persist under drought conditions. The stability that is engendered in this process acts to maintain soil quality and should allow non-weedy species to migrate between ecosystems as the climate changes. But one disturbing factor which may undermine this transitional stability is the poor state of health of forests in Europe, where pollution and acid rain, together with chronic pest infestations, have damaged the above-ground part of the partnership to the extent that the soil flora, and ultimately the forest ecosystem as a whole, may well collapse.

Most of Perry's research focuses on forests and he sees several important steps that should be taken to help to maintain the stability of woodland ecosystems and their soils during the period of rapid climate change. Management practices in particular will need to be relaxed, especially with respect to removing the understorey trees and shrubs which have no commercial value. Uniform, intensively maintained monocultures of trees are especially vulnerable and it may be the associated, incidental flora of trees and shrubs, which intensive forestry takes such pains to remove, that maintain the health of the soil flora between periods of felling and planting.

Without this bridge of mycorrhizal 'nurse' species the soils may die, especially during periods of intensive stress imposed by climate extremes. Prudent management would dictate that woodlands and forests should have a healthy covering of trees and shrubs at all times and that bare soil should be avoided at all costs. Similar arguments apply to agricultural ecosystems.

EIGHT

An Ill Wind?

> A horizontal tree – alive or dead – is at least as good a habitat as an upright one.
> OLIVER RACKHAM,
> *Trees and Woodland in the British Landscape*

The whitethroat's nest that I found in the stool of a coppiced sweet chestnut is the subject of the first entry in a natural history diary which I have kept, more or less continuously, since the third week of May, 1961.

The nest of woven grass and odds and ends of hair and wool was tucked in amongst bramble stems and a patch of wood anemones that had taken root in the stump. It contained a cluster of five greenish-grey blotched eggs, beautifully smooth and still warm to the touch.

It was more than just birds' nests which made the coppice an irresistible place to trespass in. From the moment when I slipped into it, over a ditch and through a rusty barbed-wire fence, the day took on an air of expectation and excitement.

The coppiced stems had grown upward but still let broad shafts of sunlight through to the woodland floor, which lit up patches of bluebells like flickering searchlights. Everything seemed to flicker. The young, yellow-green leaves constantly moved in the breeze, and hover flies which hung in shafts of light amongst the branches suddenly disappeared into the shadows. Sometimes the shadow of a bird passing overhead would flicker from tree to tree. Everything moved in a way that was mysterious and vaguely threatening; in a way that heightened the senses and made me acutely conscious of being in forbidden territory. Always there was the fear that one of the fleeting shadows might turn onto something real or that I might be caught in a dazzling shaft of sunlight and be spotted by a gamekeeper.

Further into the coppice the trees abruptly disappeared, cut down to the ground. Here there were piles of poles, some split for palings, stacked amongst drifts of bluebells. This was a rare example of a working coppice, but it was already falling into neglect on the day that I found the whitethroat's nest. In later years when I returned, things looked worse each time: the coppice poles grew taller and thicker, closing the canopy and shading out all sunlight. The bluebells seemed to thin out every year. Worst of all, the deepening shade robbed the wood of its movement and mystery; it became still, damp and cold.

The last time I returned was in May 1989. The trees had gone. In their place was a solid sea of bluebells. When I got out of the car and walked through them, the site of the woodland seemed much smaller than I had remembered it – hardly bigger than a large field.

It was only when I noticed that some of the older standard oaks had been snapped in half that I realized that the Great Gale of October 1987 had brought about this transformation. Tall piles of freshly sawn timber stood amongst the bluebells and all the coppice had been neatly cut back down to ground level again and was beginning to sprout. The destruction that the wind had left behind had provided the impetus for someone to manage the wood again and renew the cycle of coppicing. The very stump that hid the whitethroat's nest must have been somewhere in amongst the hundreds of recently sawn sweet chestnut stools in the bluebells.

Coppices, like all woodlands, are at their best when they are carefully managed. As mature timber is harvested and trees are constantly replanted, a patchwork of age-classes of trees develops, so that there are always some areas in deep shade and others where sunlight can reach the woodland floor. The end result is a shifting mosaic of habitats, which steadily change as the trees grow but never completely disappear.

AN ILL WIND?

Such woodlands harbour the greatest diversity of plants and animals.

This form of management is totally artificial, but natural forces can create a similar effect. Wind, rather than axes and chain saws, is the natural agency responsible for felling trees.

If you should ever feel the need to put yourself in touch with the forces of nature, I would recommend walking through woodland at the height of a severe gale. It is truly exhilarating. Some would say that this is irresponsible advice, for to do so invites being crushed to death by a falling tree. But the same people would probably not advise you against a road journey along the A1, where the chances of being killed or maimed are far greater and the spiritual rewards are infinitely less.

I happened to be in an old woodland of oak, beech and assorted other trees on the morning after the Great Gale, on one of my regular moss-collecting trips. It was an awesome experience. Overhead, the wind shrieked and howled through branches silhouetted against racing clouds. As the treetops swayed, branches clashed together and broke, tumbling to the ground in a shower of twigs, crashing into the leaf litter.

A sickly, shadowless light filtered through the remaining yellowing autumn foliage on to the thick carpet of fallen leaves. In contrast to the violence overhead, there was a relative stillness on the sheltered forest floor. Eerie little wind devils picked up small groups of bronzed leaves, whorling them in a vortex that zigzagged between the trunks and then dropping them. By leaning against the tree trunks I could feel the force of the gale bending them until they creaked and groaned. I hoped that one might snap or be ripped from the ground, but none yielded to the gale in my particular patch of woodland. A few were wounded, with branches ripped off, leaving yellow, splintered scars, but all remained standing. I was on the fringe of the storm and by then the wind was falling.

The Great Gale of the night of 15th October 1987 felled an estimated fifteen million trees – more trees than would be cut down in Britain in a whole year of commercial forestry operations. The press described it as the greatest natural disaster in England this century.

Reaction to the gale came in several forms: shock and disbelief that a climatic catastrophe that we usually associate with the tropics, and usually only experience through the medium of television, should actually arrive on our shores; horror at the loss of life; and anger that meteorologists had not only failed to forecast it, but had actually refuted warnings that it would hit England.

When the dead had been buried, the insurance claims settled and the fallen trees cut up and cleared away, there was time to reflect on the wider issues raised by this extreme climatic event. Was it a one-off, or could it be part of a changing trend in climate? If hurricanes were to become a feature of Britain's weather, what would be the economic and environmental implications?

Since the Great Gale we have had another, in January 1990, which was almost as destructive. It felled yet more trees, killed yet more people and this time it coincided with a high tide, flooding a whole town, at Towyn in Wales.

Every time a gale strikes, we count the cost in human terms but usually ignore the biological consequences of these extreme weather conditions for our wildlife. Five years after the Great Gale it is now possible to appreciate that the trail of destruction set in motion a train of events which temporarily enriched forests as wildlife habitats.

A typical, deciduous woodland is a highly organized community of plants and animals which has reached a dynamic equilibrium. Ecologists refer to it as a climax community: one that has reached the end of a sequence of changes that began when lichens grew on the boulders left by retreating glaciers and started to convert them into organic soil. Before man arrived, large areas of Britain were extensively covered with woodland climax communities.

The overlapping branches of the mature trees form a canopy which only allows a limited amount of sunlight to filter through. Below this, a shrub layer composed of species like elder, brambles, honeysuckle and hawthorn, interspersed with saplings from the forest trees, grows up towards the light. Under the shrubs and in the spaces between them grow the herbs, like the bluebells, primroses, enchanter's nightshade, wood anemones and wood sorrel, which provide much of the colour in a woodland in spring. And close to the soil surface lie the mosses and liverworts, secure in the moist shade from the drying winds of more exposed habitats.

When wind fells a tree it is not an end, but a beginning. It signals the start of a race. Young saplings grow towards the gaping window of opportunity in the canopy that the death of a fully-grown tree creates. For the first time in years sunlight streams through to the forest floor. Brambles and honeysuckle, which flower poorly in shade, bloom profusely. The woodland carpet of flowers produces a spectacular display of colour, attracting insects, which in turn become food for birds like blackcaps, whitethroats and chiffchaffs.

After the flowers come the fruits and seeds, providing a food source for birds and mammals. For a brief period, until saplings grow into trees and the canopy closes again, the flowering of the forest floor increases its biological diversity and recreates the earlier stages of the development of a mature forest.

Meanwhile, the fallen timber begins to rot. The first invaders of the dead wood are fungi, whose creeping hyphae are everywhere on the damp forest floor. They dissolve away the nutrients below the bark, reducing wood to the texture of a sponge and eventually spawning troops of toadstools and bracket fungi.

As the wood rots, insects and tiny invertebrates move into the crumbling fibres. About 20 percent of our insects and 30 percent of our birds rely on dead wood to some degree at some stage in their life cycle. Woodlice, centipedes, millipedes, slugs, spiders, larvae of wood-boring beetles and earwigs shelter below the bark, eventually attracting hungry woodpeckers in search of a meal. What was once a sound branch becomes a soggy home to the miniature beasts of the forest, sheltering below a creeping blanket of wet moss and algae.

Even the hollows left by roots which are ripped out of the ground as trees topple over are valuable micro-habitats, filling with temporary pools of water during heavy rain and harbouring their own complement of plants and animals.

Wherever woodlands were devastated by the Great Gale, the ground flora blossomed. Any dead wood that was left to return to earth via the natural cycle of decay became home to a multitude of animals. With hindsight, many biologists now believe that the great clear-up after the devastation was pursued with unnecessary haste and zeal. Had more dead wood been left to lie for longer, the literal windfall would have been even more beneficial for wildlife. A gale may well be an ill wind, but it does woodland wildlife a power of good.

Far from being a disaster, the aftermath of a severe gale is an integral part of the natural cycle of a woodland ecosystem. For the wildwood, before man began to burn and clear it for agriculture, it was the means by which biological diversity was maintained. Just as coppicing lets in sunlight and maintains the woodland flora and fauna at its best, so gales temporarily lift the sombre veil that falls over a wood when the tree crowns interlock and drown the ground flora in gloom and shadow.

Trees blowing down is as natural a phenomenon as you will find anywhere in our islands.

There are, however, some good reasons why nature should not be given a completely free hand after a gale.

In Weardale, near my home, there is a small beech wood, perched on a steep slope, which was severely damaged by the gales in the winter of 1990. Elms that had once flourished in the wood have all but disappeared, destroyed by Dutch elm disease, and the high winds that tore through the trees toppled several large and decrepit beeches, throwing open a once shady wood to an army of invaders. By the spring of 1990 the woodland floor below the gaps in the canopy was carpeted with seedlings, but few of them were beeches. The indigenous trees were outnumbered a thousandfold by sycamores, whose winged seeds were blown into the woodland and are now poised to supersede the beeches and elms.

It is hard to believe that sycamore was once a rare tree in Britain, or that it may only have been here for around five hundred years. It is a native of the mountains of central Europe and its recent origin in Britain is suggested by the fact that the sixteenth-century herbalists, who had something to say about every plant of potential or actual economic importance, hardly mention it. Once it became well established its value was soon recognized. It produces a lovely white wood that was once used for making buckets and equipment for dairies. It is still rated as a good timber tree, with fast growth and great hardiness. But it has one major fault. It enjoys life here too much. Each tree produces thousands of winged seeds, which spin like helicopter rotor blades as they drop from the branches. This slows down their fall to earth, so that they can be carried considerable distances on the wind. With its spinning seeds, sycamore spread to most parts of the British Isles in the space of only four hundred years. It has become a major ecological problem.

Once sycamore establishes in grassland, it generates scrub and then woodland in a remarkably short space of time. The dense canopy of foliage creates heavy shade, and what was once species-rich turf, studded with orchids and other desirable wild flowers, can be transformed into a relatively sterile woodland floor within the space of fifteen years or less.

This might not be so bad if sycamore acted as a host for more insects, but it has a notably poor fauna associated with it. During the brief period in which it has conquered our landscape it has only acquired a retinue of less than thirty insect pests. In contrast, birch supports about two hundred and eighty species of insects and oak hosts some three hundred and twenty.

The most notable organisms that are associated with sycamore are the tar-spot fungus and aphids. Tar-spot fungus disfigures the leaves, but otherwise does little harm to the tree. Ironically, removal of some air pollutants may actually intensify attacks by this fungus, since it is sensitive to acidity from sulphur dioxide pollution. But there is no sign that the tree itself will succumb from a greater number of black blotches on the leaves.

Likewise, the millions of aphids that cover the undersides of the leaves tap its sap and drain its energy, but they do not threaten the life of their host. The intensity of sycamore aphid infestations can be phenomenal, as anyone who has parked a car under the tree in summer will know to their cost. The minute sap-sucking insects release a constant rain of sugary drops, but they do little to limit the tree's rampant growth and aggressive spread. Like many aliens, when sycamore was introduced it left its natural pathogens behind and has yet to acquire a new retinue of pests and diseases.

So when sycamore replaces our felled native woodland trees, the transformation is certainly a retrograde one. In my local beech woodland the mature trees formed an even-aged stand and no effort had been made to nurture the next generation of saplings by protecting them from grazing animals. Unless the woodland is managed, to encourage the beech seedlings and control the sycamores, the outlook is not encouraging.

After a gale, a natural woodland needs sound management in order to maintain and encourage regeneration of its original complement of tree species. But commercial woodlands need much more intense management, to minimize the possibility of the worst of all possible forest disasters, fire.

The emergence of storm-force gales as a regular feature of winters in southern England would have a profound influence on forestry in the region. The loss of timber through windthrow, the foresters' term for the toppling of trees by wind, could render forestry uneconomic. If this climatic extreme was coupled with summer droughts, which would increase the risk of fire, growing trees as a crop might be totally uneconomic in the densely populated southern part of Britain.

Professional foresters spend a great deal of time, effort and money in removing dead wood from woodlands after gales. A thick layer of combustible material invites a forest fire, the greatest catastrophe that can befall a forest and the wildlife that it shelters.

The combination of wind and drought creates conflicting priorities in woodland management. From the conservationists point of view, leaving the dead wood to rot provides ideal habitats for some groups of animals which have declined in recent years, like the longhorn beetles that specialize in exploiting decaying timber. From the professional forester's point of view, this enthusiasm for dead wood amounts to playing with fire.

One of the worst of modern-day myths about the countryside is that all broad-leaved woods are good and all coniferous woodlands are bad. Those who preach this kind of prejudice can never have visited a well planned and managed mature coniferous forest, where several cycles of planting and felling have taken place, creating a mosaic of age-classes of trees. To my mind these coniferous woodlands are often far more biologically interesting than many beech woods.

My local evergreen forest, at Hamsterley in County Durham, is managed by as sympathetic a group of foresters as you could wish to meet, and their work has enriched the local countryside. Over the years Gordon Simpson and Brian Walker and their colleagues have painstakingly accumulated new records of rare plants and animals and taken great care to conserve them. Their forest is now a breeding site for birds of prey which are so rare that their names cannot be mentioned, and contains meadows which have scores of rare plants.

Hamsterley is one of the fast-dwindling strongholds of the red squirrel, and until recently it was free of the grey squirrels that have displaced the native species almost everywhere. This engaging rodent has steadily pushed our native red squirrel into retreat and now has exclusive tenancy of large areas of its former territory. There are a few relict populations of red squirrel in the south, in the Isle of Wight for example, but for the most part they are making a last stand in the conifer forests of north-east England and Scotland, where they are just about holding their own against the invader from the New World. The prime habitat for reds is mature Scots pine forest, during the mid-phase of the tree's life cycle, when cone-bearing is reliable.

It is uncertain how a changing climate will affect the competitive balance between red and grey squirrels, but past experience with this adaptable pest suggests that there is cause for concern. An important factor is likely to be the renewal of interest in broad-leaved forests and the mixing of broad-leaves and conifers, which tends to favour greys strongly. The vast community forests which are now planned for the north-east, composed largely of broad-leaved trees, will cause a massive population explosion of the alien species in red squirrel territory, especially if climate amelioration brings the milder winters which seem to favour the grey invaders.

What will worry foresters most will be the potential for faster growth in grey squirrel populations, since they will find that current levels of damage to trees like beech will become correspondingly worse. Large populations of grey squirrels injure trees directly by stripping bark and killing shoots and branches. This can often be fatal, since removal of bark opens the way for fungal pathogens.

Squirrel numbers usually build up after mild winters and are particularly favoured by good seed crops. This allows them to breed early, starting in January or even December, so that young are born in early spring. These new recruits to the population cause a great deal of damage and are followed by a second litter in summer.

Bumper seed crops, or 'mast years', do seem to have been a feature of recent mild winters in many parts of Britain. This could be due to a variety of factors, of which the most likely are good weather conditions during flower formation and flower pollination. Whatever the detailed causes, this abundance of winter food and good conditions for survival suggests that grey squirrel populations in a warmer Britain will be more secure than ever.

*

AN ILL WIND?

It was a bleak January afternoon and a gale was blowing, clashing together the tops of the tall conifers on either side of the forest track. I was walking head down into the wind, hurrying back to the shelter of my car. The short-tailed field vole could not have been looking where he was going either. He almost collided with my boot as we passed in opposite directions. I had taken several more steps before the peculiar nature of the encounter dawned on me and I turned to watch the tiny mammal disappearing down the track like a runaway clockwork toy, occasionally veering into the verge to investigate a possible food item.

Field voles are one of the native mammals that should benefit from climate amelioration, and during winters we can expect to encounter more animals like the one that I passed, searching for food. Not only will they be active and feed throughout winter, they will also breed more prolifically.

There are well-documented accounts of vole plagues following a series of mild winters, that have resulted in a trail of destruction. Two of the most famous vole population explosions occurred in 1813 and 1814 in the New Forest and in the Forest of Dean. While in Europe the remnants of Napoleon's regiments straggled back from Moscow, armies of furry voles were massing and threatened to destroy every young tree in two of England's major forests, stripping off all the bark and young shoots. The voles were at it again in Scotland in the 1880s, when voles ate so much grass that sheep died of starvation, and in 1891-92, when their numbers reached lemming-like proportions and enormous damage was done.

These cyclical small mammal population explosions have waned in the present century, probably because of changing agricultural practices, but they are still regular occurrences in central and eastern Europe. Population numbers may well increase here again in a milder climate. These tiny rodents thrive in young conifer plantations underseeded with dense, rough grassland, where they create mazes of temporary tunnels through the mat of dead vegetation in winter.

Small mammal population explosions may be unwelcome

to foresters, but they boost the numbers of their natural predators, like short-eared owls and other birds of prey, together with weasels, stoats and a wide variety of other birds and animals that are partial to the occasional vole. Influxes of short-eared owls are a regular feature of vole plagues; they find the superabundance of small, furry food items absolutely irresistible.

Timing of the breeding of many small mammals, like short-tailed field voles, seems to be controlled by daylength, so that lengthening days of late winter stimulate breeding and shortening days are inhibitory. Given suitable conditions, voles can breed in any season of the year, but a longer breeding season in a warmer Britain may be less significant than enhanced winter survival. Like most small mammals, the short, frantic life of the vole is measured in weeks or even days, and during winter the population normally plummets. If an improved winter climate lengthens the average life span, perhaps by only a matter of days, and the accompanying improved supply of winter plant food results in more adults arriving at the spring breeding period in good breeding condition, so producing larger litters, the impact on total populations could be enormous. Owls will have a field day.

NINE

Out of the Flames

> To many persons this Egdon was a place which had slipped out of its century generations ago, to intrude as an uncouth object into this. It was an obsolete thing, and few cared to study it. How could this be otherwise in the days of square fields, plashed hedges, and meadows watered in a plan so rectangular that on a fine day they look like silver gridirons? The farmer, in his ride, who could smile at artificial grasses, look with solicitude at the coming corn, and sigh with sadness at the fly-eaten turnips, bestowed upon the distant upland of heath nothing better than a frown.
> THOMAS HARDY, *The Return of the Native*

You can tell when spring has arrived on moorlands by the smell and the sounds. On the first warm, still, sunny day, usually in early April, the wet acid soils give off a subtle, sweet aroma of peat and dried thyme. At the same time the long, bubbling calls of curlews on their courtship flights echo across wide open expanses of heather and bracken.

Average temperatures drop by about 1°C for every 100 metres' rise above sea level, so the breeding season in these high places begins late and finishes early. Within a remarkably short space of time a host of birds arrive: some, like wheatears and ring ousels, as migrants from overseas; others, like black-headed gulls and curlews, as residents returning to their regular breeding grounds.

Britain's uplands could be described as climatic islands, where the breeding and growing seasons are compressed in relation to the warmer lowlands that surround them. It is on this higher ground that effects of climate amelioration will be particularly pronounced.

The harsh climate keeps the lowland flora at bay but, as average temperatures rise, the upland seasons will lengthen and less hardy plants will be able to migrate uphill. Invasive trees like birch will creep higher, colonizing high pastures and heather moorland, while some highly adapted animals will find themselves under pressure from their counterparts from lower altitudes. Meanwhile, the climatic range and length of growing season of bracken, which is perhaps Britain's worst toxic weed, will expand.

One of the upland mammals which will almost certainly suffer from a milder climate will be the mountain hare, a rare species which is restricted to a few cold, high moorlands. In effect these animals, which turn white in winter, are marooned on cold islands of high ground which will shrink as temperatures rise. The common hare of the lowlands, which is a declining species and is anything but common through much of Britain, might benefit from a milder climate and may be able to compete with its upland cousin for the lower fringes of the specialized moorland habitat.

Heather moorlands are degenerate ecosystems which usually develop on the sites of a cleared forest. Typically they occupy a zone between 250 and 700 metres above sea level, where winters are long and cold, frosts can persist into June and where rainfall is high. Despite these harsh conditions they are wonderful habitats for walkers and wildlife. With open views that extend to the horizon in every direction and clouds scudding across the sky, these open spaces provide a dizzy sense of freedom which is almost impossible to experience anywhere else in our islands.

In summer there are brilliant green tiger beetles that half fly, half run one step ahead on the footpaths, adders that bask in sunny patches between heather bushes, and rakishly plumaged golden plovers. There is always the possibility of catching a glimpse of rare birds of prey, like merlins that nest on the widest expanses of moorland, or hen harriers that flap low over the vegetation in search of food.

This unique ecosystem depends for its continued existence on deliberate fires, which are skillfully used to maintain a patchwork of different aged stands of heather moorland, but at the same time it is threatened by accidental fires.

Small patches of heather moorlands are deliberately burned when the peat is wet. A brief, intense blaze is allowed to sweep over old heather and dead stems, destroying the above-ground vegetation but leaving the root systems and a bank of seeds in the soil generally unharmed. Within a season regrowth begins, generating short, succulent shoots that are a major component of the diet of grouse.

A typical, well-managed grouse moor is a mosaic of patches in various stages of recovery after a sequence of burns, which are usually practised on a ten- to fifteen-year cycle. The species composition that regenerates after a controlled heather fire is virtually the same as the vegetation that existed before the burn.

Hot summers are damaging, inasmuch as they slow the regrowth of young heather after controlled burning. They can be disastrous if a fire starts during a long period of drought. One such blaze, the worst for twenty years, struck a lonely stretch of Witton Moor near Leyburn in North Yorkshire in April 1990. Moorkeeper Tom Spencely noticed a small column of smoke, but by the time the fire brigade arrived they were greeted with a wall of flame. It burned for days, despite the best efforts of over a hundred firemen. The cause was probably a discarded cigarette end, which led to the devastation of five square miles of moorland.

Later in the year a similar blaze, this time caused by a camping stove, forced the closure of Yorkshire's famous Lyke Wake Walk. Summer droughts present a major threat to heather moorland.

During a drought, peat dries out and assumes the properties of tinder. Whereas wet peat shields roots of deliberately burned moorland from heat, dry peat produces a deep-seated fire which is extraordinarily difficult to extinguish. It can burn down to the bedrock, reaching temperatures in excess of 800°C. Even when the flames have subsided, the peat smoulders for days and can burst into flame again as winds fan the surface. Such a fire destroys all vestiges of life in the affected moor. Natural recolonization must start completely afresh, on a bare, sterile surface.

The fire itself is only the first episode in the ecological disaster. Stripped of its vegetation cover, the ash and burned peatland is eroded by wind, rain and frost, leaving some areas as little more than shattered rock. It returns the land to something like the appearance that it would have had when the glaciers retreated. Life must begin again.

Soon after the 1990 fires subsided, a group of scientists from Exeter University published the results of a long-term study of another fire in North Yorkshire, which blazed over six hundred hectares of moorland during the drought of 1976. During that year there were sixty-two uncontrolled fires on the North York Moors and those at Rosedale Head were amongst the worst.

Ten years after the blaze the Exeter research team found that large charred areas devoid of vegetation still dominated the landscape. Heavy erosion had scarred the land surface. Mosses, arriving as wind-blown spores which germinated on the charred surface, dominated large areas of the land surface, but the vegetation that had existed before the fire had barely recovered at all. Such is the instability of the burned site and so harsh is the upland climate that a recognizable recovery of the moorland will take decades.

The decline of the large moorland shooting estates has almost certainly increased the risk of uncontrolled and deeply damaging fires in drought years. The traditional management practice of regular burning limited the amount of above-ground fuel that could accumulate, so accidental fires were starved of combustible material. Now that many landowners can no longer afford such intensive management, extensive areas of heather have become overgrown, creating a large initial fuel source for an accidental fire.

Whatever the arguments for and against grouse shooting, there seems to be little doubt that traditional management practices maintained the much-loved heather moorland in a productive state, which supported a unique community of plants and animals, ranging from emperor moths to merlins.

If a warmer climate brings summers of the kind that we have experienced in recent years, there will be many more moorland blazes. When fire risk is high, measures will need to be taken every year to reduce public access, minimizing the chances of disastrous fires ignited by cigarettes and camping stoves.

We tend to think of much of our landscape as being natural, and yet very little of it is unmanaged. The species composition of deciduous woodlands needs to be controlled to maintain their continuity. The traditional maintenance of heather moorlands needs to be continued in order to safeguard it from disastrous fires in a drying climate.

If there is one certain prediction that can be made about our response to the changing weather pattern, it is that we must be prepared to manage the environment even more intensively than we do at present, in order to ensure the survival of the landscape features that we traditionally value most. The concept of a natural landscape will have even less meaning in the twenty-first century than it has at present.

Lowland heaths are yet another habitat created by the human hand. About five thousand years ago Neolithic hunters began to clear the forest, probably using the same kind of slash and burn techniques that are employed by primitive cultures in tropical rain forests today. Once the land was cleared, they briefly cultivated it until the soil fertility was exhausted, then moved on to burn and cultivate new land.

Despite its lack of soil fertility, the abandoned land would have progressed through a sequence of vegetation types, from grassland, through scrub to woodland, had it not been for large herds of grazing animals, like red deer, which suppressed tree growth and maintained an open habitat of the type that we now associate with lowland heaths, like those in Thomas

Hardy's Dorset or in the Brecklands of East Anglia. Some tree species, notably birch, are adept at recolonizing heathland and once established are hard to control, but the combination of grazing and occasional burning would also have kept the development of woodland at bay. As the large herds of grazing deer disappeared from the countryside, the introduced rabbit took over and maintained the short vegetation, until it too was temporarily exterminated by myxomatosis.

So heathland survived over the centuries, often as common land; but by the eighteenth century it began to decline and in the present century it has rapidly disappeared.

Now, lowland heaths are amongst the most precious and threatened of all semi-natural wildlife habitats, and in hot summers, like that of 1976, accidental fires have often destroyed large areas. Where sufficient surrounding heathland remains, this may not be as disastrous as it might first appear. After all, heathlands have been created and maintained by fire and constant disturbance and this temporary setback allows species to recolonize and the heathland habitat to reform. But where small pockets of heathland, isolated between built-up areas and large tracts of agricultural land, are destroyed, recolonization is impossible and the damage is long term and irreparable.

The southern heathlands have much in common with heather moorland in terms of some of the plant species that they support. Heather is a characteristic of both. But whereas upland moors are generally wet and typified by the development of peat, the lowland heaths are dry, often develop on sandy soils and support a unique assemblage of animals that thrive in sun-drenched habitats. Reptiles flourish particularly well in such places, where low-growing vegetation provides cover but does not shield them from the sun.

When I was a child, reptiles held a particular fascination. Fortunately I lived in a part of rural Sussex where slowworms, common lizards and grass snakes were still fairly plentiful, and I soon learned that if you wanted to catch them, you needed to be stealthy. Slowworms and lizards lived happily at one end of my grandmother's allotment. The trick was to creep up the bank without snagging your clothes on the brambles, gently curl your fingers under the sheet of rusty, corrugated iron and lift it quickly, with one smooth motion.

Most days, there would be two or three slowworms beneath; if you were really lucky, a lizard. The mercurial lizard would disappear in a flash, but the slowworms were easy to catch. They curled through your fingers, sinuous, muscular and smooth. Like live metal. They felt cool, even though they had spent the afternoon baking under the orange-brown sheet of metal. Their elegant eyelids blinked in the bright sunshine and their tongues darted like tiny flames.

Catching the reptiles that basked beneath the rusty corrugated iron under the overgrown hedge in the allotment, or amongst the sand dunes at West Wittering in Chichester Harbour, was a favourite pastime in school holidays. But it was a waste of time hunting them on dull, cold days. Snakes and lizards need warmth.

Sunbathing is something that we do for pleasure. For reptiles it is an absolute necessity, fraught with risks. Put yourself in a lizard's skin for a moment. Your blood is cold and until it warms up you feel sluggish and torpid, barely able to move. After a cold night you need to sunbathe to raise your body temperature, a hazardous business in the early stages, until you have soaked up enough heat to sharpen your reflexes and loosen your muscles. Even digesting a meal is a problem in cold weather, and you need to seek out a sunny, sheltered spot to allow your digestive enzymes to reach optimum operating temperature. And if you are a female you need to spend hours in the sun, so that your eggs can complete their development before you lay them.

Provided that their other habitat requirements are satisfied, reptiles should benefit from warmer summers. Just how important sunshine is to lizards can be seen from studies carried out at Southampton University by Dr Ian Spellerberg and his collaborators. Using a seasonal measurement of sunlight known as the sunshine index, he has compared the solar heating requirements of the endangered sand lizard with those of the commoner and more widespread viviparous lizard. Sand lizards seem to favour areas with at least twice as much sunshine as their commoner cousins, and their newly-hatched young are even more particular about where they sunbathe.

Sand lizards lay eggs which are abandoned by the parents, and the hatchlings need a very high sunshine index. Viviparous lizards give birth to live young and may be able to use their own basking behaviour to the benefit of the developing embryos inside their bodies.

In the case of both species, sunnier summers are certain to be beneficial, and should help to increase the populations and perhaps the range of the common viviparous lizard.

The period between May and October 1989 would have suited reptiles well. This was the fourth warmest since the Central England Temperature record began in 1659, with an average temperature 1.3°C above the 1961-80 reference period. The summer of 1989 was the sunniest since 1909. More summers like that might well help our reptile population, but for the sand lizard they may arrive too late.

These disappearing reptiles are mainly confined to heathlands in Hampshire, Surrey, Sussex and Dorset, with one small outlying population on the Merseyside coast. The warmest climate in Britain since the last glaciation persisted about five thousand years ago, and since then the weather has gradually deteriorated, so the main centres for sand lizard populations have retreated steadily southwards. Except for the Lancashire lizards. They clung on in the north, where average annual temperatures are over a degree lower than in Dorset, for five thousand years.

The reasons for the persistence of the curiously isolated sand lizard colony on the sand dunes and golf courses of Southport remained a mystery until Helen Jackson at Manchester University's Department of Zoology examined the isohels of the southern and northern sites. Isohels are lines drawn on weather maps which link up places that have the same amounts of sunlight. She found that the $6\frac{1}{2}$ hour May isohel – the line joining places which have a daily average of $6\frac{1}{2}$ hours of bright sunshine – encompassed the southern populations, skirted the edge of the Welsh coast and took in the narrow sliver of Lancashire coast that hosted the sand lizard colonies. Clearly, the duration of sunshine on bright May days, when the lizards emerge from hibernation, was the key factor in their survival in the north-west.

It seems certain that one essential requirement for reptiles' survival is plenty of spring sunshine. This coincidence of sunshine hours and animal distribution clearly highlights a frustrating problem for those who like to attempt to predict specific effects of climate amelioration; for many species the precise seasonal timing of weather changes might turn out to be far more important than changes in mean temperature or hours of sunlight or inches of rain.

May was a poor month for sunshine in 1976, a year which gave us the hottest summer for two hundred and fifty years. The average daily hours of May sunshine on the Lancashire coast during that memorable year dropped from a blood-warming 6.6 hours to a numbing 4.4 hours. Despite a sizzling summer, it was not a good year for the Lancashire lizards.

No one knows exactly how the cold spring of 1976 affected this isolated lizard population, but it was probably less damaging than man-made changes. These animals have more precise habitat requirements than most British reptiles, and their survival also depends on the presence of the right kind of vegetation, which is rapidly disappearing as many of the favoured heathland areas are converted to agricultural and building land or are used for forestry. Much of the Lancashire colony has already been destroyed by development.

So whatever the future weather, preservation and extension of existing reptile habitats will be critical, irrespective of how favourable the new temperature regime will be to the lizards themselves. A warmer climate is not a ticket to survival, even if the distribution of a species does depend on temperature and sunshine.

TEN

The Teeming Hordes

> Since the beginning of the nineteenth century the English countryside has become steadily less favourable for our butterflies. Fens have been drained, waste land reclaimed, and urbanization and industrialism have made vast and hideous advances ... The extent to which our fauna and flora have gained through the establishment of nature reserves ... has been negligible in comparison with such evils.
>
> PROFESSOR E.B. FORD, *Butterflies*

Our modern, industrialized society has done much to degrade and erode habitats for wildlife, but there is one group of animals that has consistently benefited from the changes that have unfolded during the twentieth century. Our relentless search for an improved standard of living has also improved the quality of life for the insects that have become adapted to live in close association with us, our crops and our animals. For them, we are constantly creating new and improved habitats, and as the temperature rises they will thrive as never before.

If we named our years after animals, according to the Chinese system, 1990 might go down as the Year of the Cat Flea. During the summer of that year 1360 letters flooded into the Medical Entomology Centre at Cambridge University, as anxious cat lovers sought advice on how to deal with a huge jump in the flea population. Three years of warm winters and hot summers had simply caused a flea population explosion, with larvae and pupae developing more successfully in warm, damp cat fur in mild winters. Come summer, the higher temperature accelerated flea development, allowing time for a few extra generations over the three-year period.

An average cat carries about ten thousand fleas, mostly as eggs or larvae. Most cats carry about a hundred adults, which tend to commute between pussy and the floor covering. Deep-pile carpets are just one of the many improvements in our standard of living that have provided new habitats for wildlife. By the summer of 1990, pet lovers began to become seriously concerned about the build-up of a flea fauna in their best Axminsters. Sales of methoprene spray, used to rid moggies of this irritating pest, rocketed.

During the same summer wasp numbers soared too. Wasps are much-maligned and misunderstood insects. We have much to thank them for. A moderate-sized wasp colony will rear around fifteen thousand individuals in a season, all fed on caterpillars, flies, greenfly and similar garden pests, which are hunted by foraging workers.

I once witnessed a particularly gruesome, one-sided combat between an orange tip butterfly caterpillar and a common wasp. As the wasp homed in on the hapless caterpillar, its victim twisted and turned, desperately rearing up to drive the attacker away. Finally the wasp landed on the caterpillar's back, secured a grip on the segment behind the head and set about chewing it off. This took some time; the larva was fat and the wasp's jaws were small. Eventually the wasp flew off, carrying the head back to the nest. The headless caterpillar continued to wriggle and rear up, still clinging to the plant, until the wasp came back for another chunk. Such incidents are an everyday occurrence in the garden, and during the course of a season a wasp colony will account for thousands of pests.

It is only in autumn, when the wasps' nest begins to break up, that they become troublesome. No longer required to hunt food for the brood, the workers attend to their own needs and are attracted to anything sweet, entering houses and generally causing panic. Many end up squashed under a rolled newspaper; of those that last until the first frost only the queens survive, hibernating in lofts, outhouses and under tree bark.

Recent hot summers have done wonders for wasps. As is the case with all cold-blooded animals, everything happens faster when the temperature rises. Eggs and larvae mature more quickly, colonies are founded earlier and numbers of offspring increase rapidly. Modern houses provide excellent sites for wasps' nests, in warm, insulated loft spaces or below floors. So with a good summer and plenty of secure habitats to build nests in, wasps thrived.

By August of 1990, as summer temperatures hit an all-time high, wasp numbers reached colossal proportions. Between May and August 1990, Barnet Council's pest controllers removed 2200 nests from their area. Apollo Pest Control Ltd removed as many as 400 nests per week. Similar wasp plagues were reported in Sussex, where health officers in the Arun District received a 75 percent increase in complaints of wasp nests during July and August alone. The three local environmental health officers there could barely keep pace with the volume of complaints.

But in a few places the relentlessly hot weather also spelled disaster for some colonies, when lack of water forced them to feed on willow aphids. One of the well-known medicinal uses of willow is as a cure for headaches. Willow, in the genus *Salix*, contains high concentrations of salicylic acid, the active ingredient of aspirin, in its sap. Willow aphids tap this sap, and although they seem to be unaffected by the acid, the wasps feeding on the aphids quickly accumulate a drug overdose. Reports spoke of carpets of stupefied wasps under willow trees.

There no longer seems to be any doubt that our climate is changing. There is growing concern that the pace of change will be so rapid that many plants and animals will have insufficient time to evolve and adapt to the sudden shift in seasons. But whatever might happen to the rest of our flora and fauna as our climate changes, insects in general might be expected to cope quite well.

Theoretically, they have three key credentials which should give them a reasonable chance to overcome the worst effects of fickle weather patterns: they often reproduce in very large numbers, they usually have short life cycles and they are generally very mobile. The first two characteristics are essential for rapid evolution and the last is useful if the only means of survival is to migrate.

A living organism has two choices if it is faced with a deterioration in its environment. It can move to somewhere more favourable, or it can stay put and slowly evolve so that it becomes better adapted to the new conditions. If it can do neither of these, it will become locally extinct.

The most famous case of insects adapting to man-made environmental change occurred during the Industrial Revolution, when peppered moths actually changed colour, from pale grey to battleship grey.

The first dark-winged specimen of a peppered moth was captured amid the smoking chimneys of Manchester's industrial heartland in the 1840s, and by the end of the decade about one percent of the local individuals would have been of this so-called melanic form. By the end of the century an estimated 99 percent of the population was melanic, and the pale-winged form had become a local rarity. Even as Charles Darwin sat down to write his *Origin of Species*, these moths were undergoing a colour change which still remains one of the most striking examples of evolution in action ever witnessed by biologists.

It fell to two twentieth-century ecological geneticists, E.B. Ford and B.D.H. Kettlewell, to explain the strange case of the disappearing pale peppered moths. They were being eaten. Camouflaged against the soot and grime deposited by industry on tree trunks, dark-winged moths were less visible than the pale form, which became easy prey for predatory birds.

A number of other moths are now known to have undergone the same transformation, as have ladybirds. In some areas black ladybirds with red spots have replaced the familiar red forms with black spots. The significance of this is still controversial, but one school of thought links it with climate. The black forms absorb more heat on sunny winter days and, although they may risk overheating, their darker colours allow them to warm up faster in spring and early summer, so that they mate more frequently and leave more progeny. If this is so, they are probably already showing adaptation to climatic conditions. Changes like this strongly suggest that overwintering insects that can produce several generations per year and large numbers of offspring should be able to tap their reservoir of genetic variability and adjust to warmer winters and early springs.

For the majority of insects that can fly, a change in climate may well allow them to move to more favourable areas or even to expand their range. The current distribution of grasshopper species is very definitely centred on the south of England, since sunshine is an essential resource for their particular lifestyle.

If any sound can be said to epitomize a hot summer day it must be that of stridulating grasshoppers. The constant chirruping, produced as they drag their rough hind legs like violin bows across the edge of their membranous wings to generate a monotonous song, is redolent of afternoon picnics on dry, grassy banks scented with the aroma of crushed thyme.

Climate amelioration will be kind to British grasshoppers, which in a European context are a select band of hardier species that have adapted to our generally cool, wet climate.

Grasshoppers have a protracted life history, which begins when they hatch from overwintering eggs in late spring and then progresses through a series of nymphal stages. In common with all invertebrate animals with a hard outer skin, grasshopper nymphs must shed their exterior skeleton by moulting before their bodies can grow. This transition between nymphal stages, or 'instars', makes development a slow, risky business, and by the time they finally achieve adulthood much of the summer has passed. The great green bush cricket, *Tettigonia viridissima*, still fairly common in southern England, plods through nine instars before it reaches maturity.

Given a decent summer, grasshoppers can move through these nymphal developmental stages smartly and reach breeding condition in time to reproduce in large numbers. An insight into the critical role played by good weather in the success of one of our rarest grasshoppers, the wart biter, has recently come from a study of this elusive insect by two scientists at Imperial College, A.J. Cherill and V.K. Brown.

To be strictly accurate, the wart biter, *Decticus verrucivorus*, is not a grasshopper at all but a closely-related bush cricket. Bush crickets can be distinguished by a long and apparently menacing egg-laying device, the scimitar-shaped ovipositor, which protrudes from their rear ends. This is harmless, but if caught the animals can deliver a painful nip with their jaws. This accounts for the wart biter's curious name, given to it by the great Swedish taxonomist Linnaeus, who claimed that his countrymen used the insect to excise their warts.

In Britain the wart biter is an endangered species, protected by the 1981 Wildlife and Countryside Act and confined to just four sites in southern England. It was a late arrival, one of the last of its family to scramble across the land bridge from the Continent before it was flooded by the English Channel eight thousand years ago. In all probability wart biters have always been uncommon animals in England, since it is a dodo among bush crickets – flightless and therefore severely limited in its ability to colonize suitable new habitats.

Here it breeds on warm, south-facing slopes. Cherrill and Brown's study has shown that eggs hatch in mid-April and that there are seven nymphal instars before adults finally emerge in July. Then the adults put on weight, finally reaching breeding condition in September. By the second week of October all adults are dead, so egg production occurs in a narrow window of a few weeks in early autumn. Bad weather in September can be disastrous, and the whole life cycle of the insect hinges on a delicate temperature balance depending on weather, aspect of slope and the structure of the vegetation.

The wart biter must have areas of short, open, south-facing turf where it can sunbathe, raising its body temperature and speeding its progress to breeding condition before low night temperatures put paid to its chances of producing eggs. Nymphs cease to develop below 20°C, and the ideal temperature for their speedy growth is around 33°C. Without sunshine it is unlikely that their body temperature could even reach 30°C. A continuing trend of long, hot summers could be the salvation of this beleaguered bush cricket, provided that a suitable mosaic of tussocky grass and close-grazed turf is maintained in its preferred habitat. As with so many of the rarer elements of our flora and fauna that could benefit from the climate change, the provision of suitable habitats is critical. Just like the sand lizard, the wart biter will need a helping hand in finding new, satisfactory habitats if it is to benefit from sunnier summers.

For other grasshoppers and their allies that retain the power of flight, sweltering summers should offer the opportunity to expand their range, quite literally into pastures new, provided that suitable pastures lie within their flight range.

Predicting the effects of warmer summers and milder, wetter winters on insect numbers is difficult because of the complex life histories of many species. A change in climate might easily favour one stage of the life history but threaten another. Hot summers might favour egg laying, but wet winters might well reduce survival of eggs or pupae, and the effect of climate change will vary from species to species. One case in point is that of the common coenagrion damselfly, which breeds in still water.

Damselflies are the ballerinas of the insect world, with slender bodies in electric hues of blue, red and green, and large muslin wings that give them a slow, fluttering flight. Watching these delicate insects swarm over a small pond at the height of summer, with eager males chasing mates, is very much like watching the complex choreography of a ballet.

David Thompson at the University of Liverpool has made a study of the effect of weather on egg laying in the common coenagrion and has shown that the production of eggs during the insect's lifetime increases in sunny summers. In a good summer, of the kind that we had in 1989, the female damselflies can produce as many as 740 eggs during their ephemeral lifespan; in a poor summer this number can drop to less than half this figure. Females that reach sexual maturity during a spell of good weather are the most fecund, so long summers with warm, sunny weather spread throughout the flight period of the adult insects are bound to increase egg-laying capacity.

But, of course, egg laying is only one component of the life history of the damselfly. The eggs hatch into nymphs, which spend the rest of their life cycle in the water. In some species nymphs moult as many as a dozen times and may take two years to metamorphose into an adult. Hot, sunny summers may increase the egg-laying capacity of adult females and may result in the emergence of more nymphs, but this will be to no avail if long droughts cause the breeding pools to dry up. This certainly happened in some pools on a nature reserve near my home in County Durham in 1989 and 1990, when shallow ponds dried to dust bowls, killing all the dragonfly and damselfly nymphs in mid-development.

So a good summer for damselflies, and probably for many other insects that spend part of their life cycle in shallow water, is one with plenty of sunny days, but with enough rain to maintain water levels in ponds.

Although many more detailed studies will be needed before it will be possible to interpret the effects of climate change on insect populations with any degree of accuracy, short-term studies have already shown that the numbers of different species fluctuate wildly in exceptionally hot summers.

John Coulson, my colleague at Durham University, spent a good deal of time in the exceptional summers of 1989 and 1990 examining the dynamics of insect populations on high moorlands, setting pitfall traps to sample the insect fauna. Although his studies only cover a short period, they do reveal dramatic reversals in the numbers of several different insect species that live in these high altitude habitats.

It seems certain that an increase in average temperature will have its greatest proportional effect in these upland areas, where the normal growth period for plants and breeding period for animals is much shorter than in the lowlands. Such habitats may be some of the best places to monitor the earliest biological effects of sustained climate change.

John Coulson's records certainly reveal a drastic change in fortune for many insects caught in his pitfall traps during abnormally hot summers. One of his most interesting observations is that the fortunes of closely related species within insect families, like the crane flies, varied drastically. For example, one species of crane fly decreased by 99 percent, whilst another – the grouse fly, an important food resource for these game birds on grouse moors – increased by 188 percent. Similar contrasting trends were also apparent in different species of rove beetles.

These studies of moorland insects, which are still at an early stage, do seem to show that climate stress will affect different species very specifically and will probably depend on such factors as the timing of the various stages of their life cycles and the effects of climate on their food plants or animal prey.

Insects species are the most numerous and diverse of all the elements of our fauna, but nature lovers are probably more concerned about the fate of butterflies than of any other insect group. The fact that they are aesthetically appealing perhaps gives them a disproportionate importance in relation to other, less attractive insects, but the shifts in their fortunes are also important indicators of the health of the environment in general.

Changes in land management, and particularly in intensive agriculture, have had a devastating effect on butterfly populations in the second half of the twentieth century. Now that the climate is changing, there is further concern for their future.

One study of Britain's butterflies suggests that new weather patterns would have a major effect on their abundance and distribution. It also highlights the same kind of species-specific effects that were evident in John Coulson's studies of moorland insects.

The survey has been conducted by two leading butterfly watchers, R.L.H Dennis from Manchester Grammar School and T.G. Shreeve from Oxford Polytechnic, and it contains a mixture of good and bad news for everyone who likes butterflies.

We tend to associate these colourful insects with bright, sunny days and there is no doubt that higher temperatures mean that they will fly more actively. Basking with their wings angled to catch the sun's rays on a cool day is a characteristic and essential part of their behaviour. On dull days they are sluggish and easy to catch; on bright days they seldom settle for more than a few seconds.

More warmth should speed their development, so that some species, like the small heath, which produces two broods in the south, will also produce additional successful broods in the north of Britain. The small tortoiseshell and the speckled wood should also respond in the same way. But for species like the pearl-bordered fritillary, producing an additional brood may well be risky, at least until the climate stabilizes. If the additional generation does not reach a developmental stage where it can overwinter successfully, in this case as a hibernating caterpillar, an additional, reckless cycle of breeding could spell disaster for the butterfly population in the following year. So additional life cycles are only of value if they are synchronized with the changing seasons.

Other direct effects of higher temperatures, like seasonal droughts or, worse, several successive summers of drought, will also adversely affect butterfly food plants and disrupt some of their more specialized habitats, like wetlands and woodlands. Hot days also reduce flower nectar production and thicken nectar, making it more difficult for some butterflies to feed.

So there are numerous factors that need to be taken into account before any predictions can be made about the future of a butterfly species.

Dennis and Shreeve have constructed an index of vulnerability of British butterflies, which takes into account factors like their current range and frequency throughout the country, their host plant specificity and abundance, the vulnerability and range of their habitats and their ability to disperse to new sites. What they have come up with is a league table of potential butterfly winners and losers in a warmer Britain.

Top of the list of beneficiaries from climate amelioration are the butterflies whose caterpillars feed on our crops, the large, small and green-veined whites. Their life styles and choice of food means that they have an assured future. Gardeners can expect to see even more of the foliage of their prize cabbages reduced to a skeleton of leaf veins. The caterpillars of these species are prodigious feeders. Once, on a still summer's day, I actually heard them demolishing one of my cauliflower plants. The thick, succulent leaf acted as a sounding board, amplifying the noise of their jaws as they chewed through the juicy tissue.

Also tipped to do well are the migrants from southern Europe, like the painted lady, red admiral and clouded yellow. At present the painted lady cannot survive hibernation in our winters, although it can breed on the very common creeping thistle. Similarly, the red admiral breeds on the ubiquitous stinging nettle but perishes in the winter. The exquisite, sulphur-coloured clouded yellow has a caterpillar which feeds on clovers and vetches, but the adults only survive our winters in very small numbers. Large numbers of clouded yellows reached our shores in the summer of 1990, and a combination of milder winters and food plants which are common everywhere will stand these immigrants in good stead when our climate changes.

In the past, avid butterfly collectors would go to almost any lengths to add the ultra-rare migrants, like the long-tailed blue and the Queen of Spain fritillary, to their collections. There are a few sightings of these each year in the south of England, but in the future these are likely to become increasingly frequent visitors. And some deliberately introduced species, like the remarkable map butterfly from central Europe, might also make themselves at home here.

The map is a woodland species that was released into Herefordshire and Monmouth in 1912 and showed signs of establishing until it was exterminated by a malevolent collector. We certainly have plenty of its food plant, the stinging nettle. It is an interesting species, not least because the generation that emerges in spring from overwintering pupae is chequered russet red and deep brown, while the second generation, which emerges in summer, is predominantly chestnut brown, with white blotches and small russet streaks. It would certainly be a welcome addition to our fauna, not least because we would be effectively gaining two butterflies at once!

One of our rarest breeding species is the Glanville fritillary which, like several of the fritillaries, looks prettiest when its wings are closed. Then, the wings are like stained glass windows, with two bands of orange, edged in black, standing out against the cream background. It maintains a tenuous foothold on the Isle of Wight, in the shelter of crumbling cliffs which act as sun traps and provide a suitable habitat for the ribwort plaintains that its caterpillars eat. In future, if hot summers become the norm, it might be able to survive on the mainland; past attempts to establish it on the northern shore of the Solent have produced short-lived colonies, one of which still survives in Somerset.

Amongst the common butterflies, the beautifully marked common blue, the small copper, the small heath, the orange tip and the green hairstreak are all tipped to increase in numbers. All of these are relatively adaptable species and tend to use food plants that grow in habitats that will be least affected by the climate change.

But Dennis and Shreeve's league table indicates that we will pay for the gain of these more exotic species and the increase in common species by losing some of our existing rarities. Those that are most at risk include the spectacular swallowtail, now confined to a few fenland sites in East Anglia.

A drier climate, coupled with a sequence of severe summer droughts, could deliver the final death blow to this and a handful of other rarities, like the large copper, which has already become extinct in this country once and has been artificially re-established from continental stock.

The native stock of large coppers was collected to extinction and the last, from Holme Fen, is reckoned to have been impaled on a pin in 1847. As the population shrank, rural workers in the Fens, who supplemented their meagre agricultural wages by selling the butterfly to collectors, watched the price for fresh specimens rise from £1 to £20, a tidy sum in the mid-nineteenth century. Our present population of large coppers at Wood Walton Fen was deliberately introduced by lepidopterists, but now teeters on the verge of extinction again, although it will probably be maintained with captive-bred specimens.

Also high on the vulnerability list are the black hairstreak, the adonis blue, the purple emperor, the silver-washed fritillary and the mountain ringlet. The latter species is restricted to grass mountain sides in the Lake District and the Highlands of Scotland, where the caterpillars feed on mat grass. The food plant is common enough on poor soils at high altitudes throughout the north, but warmer weather might well drive the insect higher into the mountains, deceasing the size of colonies. The Scotch and mountain argus are also likely to find themselves in a similar predicament.

Although species that are restricted to high altitudes may suffer, the outlook for butterflies in the north may be better than in the south, with more gains than losses. A one-degree rise in average temperature is equivalent to moving about a hundred miles south, so by the middle of the next century butterflies in Northumberland will experience the kind of temperature regimes that their cousins currently enjoy in the Midlands. Since temperature and length of season are amongst the most important factors limiting the distribution of butterflies, many species should begin to migrate northwards.

Dennis and Shreeve predict that the small skipper, which

is one of our rarer northern butterflies and which reaches its northern limit in Northumberland, should respond well to a milder climate by extending its range. Likewise, the locally rare ringlet, large skipper, small pearl-bordered fritillary and dark green fritillary should benefit from the climate shift.

The hot summer of 1990 was characterized by a spectacular population explosion of the holly blue in suburban gardens in the southern half of England, and wanderers reached as far north as Newcastle, for the first time in many years. Its food plants, holly and ivy, are plentiful everywhere, so a warmer climate may see it establishing itself as a breeding species in the north of England.

The north might also see the very welcome reappearance of the long-lost comma butterfly. This has an uncanny resemblance to a dead leaf when its wings are folded and was once very common in Northumberland and Durham, but became extinct in the north early this century. It seems to be one of the species that will be least vulnerable to future climate change, so warmer weather may speed its return. Other returnees to northern England might include the speckled wood and the grayling.

If these butterflies are to return or increase, a major priority will be to ensure that habitats are maintained that will support them. Conservationists will need to be vigilant, as a changing climate makes increasing demands on agricultural development in areas which larger populations of butterflies may migrate into.

If a drier climate forces the focus of agriculture northwards, many existing butterfly habitats in northern counties will be threatened. New butterfly arrivals will only be able to establish themselves if suitable habitats are maintained and food plants are available. At the same time there will be a need to conserve habitats which will allow many of the most attractive residents to realize their potential for responding positively to a warmer climate.

Climate change will present the greatest challenge in wildlife conservation that has ever been faced in Britain. The changing status of butterflies might represent a sensitive test of our ability to ensure the continuity of our flora and fauna and take advantages of new opportunities for increasing the abundance of some species.

But whatever happens, it seems that there will be butterfly species that are unlikely to survive the climatic upheaval. So if you want your children to see a wild, living swallowtail butterfly, best take them on a trip to the last remaining Norfolk strongholds soon. By the time they grow up, the British race of the swallowtail could have vanished completely.

ELEVEN

Nibbling Away at the Edges

> Now the sea returns to its usual leaden black, an eerie mirror for the chilled, magnificent light so essential to this landscape and its sounds – the lonely cries of gulls, peewits, curlew and, inevitably, the wailing wind. And what a wind it can be as it swipes across the punished flats; for the land is all too vulnerable to a merciless sea which fingers its way across the mud in a series of creeks. Occasionally, the wind, moon and sea join forces and then the menace of a flood is born. Not content with the sodden marsh the sea gallops up the lane and rushes through the already salt-encrusted gardens. One year it came at night, dismissed the dyke as though it were some childish seaside dam and plucked whole families from their beds. Even houses quite far inland had water up to the mantelpiece and to this day the salt has played havoc with the mortar, bricks and plaster.
>
> MICHAEL BERKELEY, *From Flint to Shale*

Having spent the first twenty years of my life living on the coast, I have a special fondness for sand dunes, salt marshes and tidal creeks.

Some people find salt marshes melancholy and desolate places; featureless expanses of mud and Spartina grass that form a no-man's-land between the civilized world of roads, farms and houses and the expanse of ocean which stretches to the horizon. At times they are, and that is one of their attractions.

On salt marshes the rhythms are tidal, dictated by the phases of the moon, and have little to do with the man-made rhythms that create the pressures of everyday life. But now it seems that these contemplative places on the edges of our islands are threatened by the gaseous effluents of industrialized society. As the temperature increases and sea level rises, the tides that form salt marshes seem likely to destroy them.

Salt marshes and the coastal waters that flood their channels are the habitats for plants and animals of exquisite beauty and complexity. For several years I kept a saltwater aquarium, stocked with animals that I caught from the network of creeks and reaches that form Chichester Harbour, which is fringed with extensive salt marshes. To catch the animals, I trawled from a small dinghy that I kept at the head of a winding, muddy channel that linked Bosham Creek with the open sea.

The best time to go trawling was on a spring tide in late afternoon. The tide would slide into the narrow creek, gurgling and bubbling into lugworm burrows, coaxing crabs out of their temporary intertidal shelters under the banks of Spartina grass and eventually lifting the rowing boat clear of the smooth, brown mud. Small shoals of silver mullet followed the tide in, glinting in the sunlight and flashing under the oars as I rowed out into the main channel.

The sea wall on the far side of the creek offered shelter where I could slip the oars and let the boat drift in the current. A heron sometimes rose ponderously from the water's edge as I passed and flew away across the harbour, while rabbits bolted into burrows where the rough grass of the field edges had been eroded by high tides.

The plankton net ballooned out astern like a submerged windsock, acting as a giant sea anchor. It would take the best part of fifteen minutes to row two hundred yards along the shore line before I paused to haul in and empty the contents of the collecting bottles. Apart from a few small, darting objects in the water, it always looked a disappointing catch, until I got it home and emptied it into the aquarium.

Almost invisible gooseberry-shaped comb jellies slowly rose and fell in the water, clear as glass, covered in rainbows of iridescent cilia and trailing long stinging tentacles. When I turned the light out they glowed with a weak green phosphorescence that gently rose and fell as they swam to the surface of the aquarium and then sank to the bottom.

Under the microscope the intricate world of coastal plankton was a revelation, populated by porcelain crab larvae with red and white jousting poles on their carapaces, tiny coelenterate medusae and prawns so transparent that you could read through them and which were only given away by their stalked eyes. The variety was endless: wriggling marine worms, fish parasites, rapacious arrow worms, and tiny mollusc larvae like living spinning tops.

Every catch would reveal new and sometimes larger wonders: red-bellied sticklebacks from the brackish head of the creek; pipefish like bootlaces with sea horse heads and vibrating fins; sea spiders composed almost entirely of legs; baby cuttlefish with tentacles that struck at lightning speed when offered a meal of sand hoppers.

But little of this was visible in the catch as I emptied it into the collecting bottles. These microscopic delights would come later. Now that the bottles were full, I could explore the creeks and inlets.

At flood tide the boats swung idly on their mooring ropes, their halliards tapping against masts in a desultory breeze. With the nets stowed I could head towards a lagoon, cut into the land by sea traders as a safe haven for unloading their ships, now a silted pool slowly engulfing the gaunt ribs of a decaying coastal sailing ship. The high tide covered the surrounding mud banks, leaving papery, pale blue heads of sea lavender and pink thrift flowers held just above the level of the water surface, luring small tortoiseshell butterflies and lumbering bumblebees into making the perilous journey from the shore in search of nectar.

As the boat glided into the sheltered water, agitated shelducks with enormous foster families often paddled past. If I splashed the water with an oar their flotilla of ducklings would dive below the surface in unison, all bobbing up again like fluffy corks a short distance away. They nested in disused rabbit burrows on the water's edge.

If I leaned over the side of the boat I could watch crabs sidle across the muddy bottom, shoals of tiny fish twist and turn in formation in the shallow, warm water and slowly pulsating *Aurelia* jellyfish drifting by.

It was easy to become mesmerized by the underwater ballet, stay too long in the lagoon and become trapped by the mud banks that were exposed by the falling tide. It was often long past sunset before I managed to thread my way out of the creeks and row back to the jetty. As the light failed, small patches of phosphorescence emitted by luminescent plankton would sometimes appear around the oars as they disturbed the water. Sometimes, as the bows of the boat slid alongside the jetty, I would disturb a redshank that had already settled for the night.

It was difficult to leave and go home once the boat was secure and the oars and the catch were unloaded. From across the salt marsh, plaintive calls of waders would float over the water and I would hear the sounds of trickles of receding water as the creeks dried out and other unidentifiable noises that came from the darkness close by.

If you care to look this closely, there is nothing desolate or lifeless about salt marshes. When the tide recedes, leaving acres of mud, birds replace the marine life, feeding on the rich supply of worms and shellfish hidden in the ooze. Arm yourself with powerful binoculars, sit on a sea wall beside a salt marsh, and you will discover that the vacant mud flats are covered with hundreds or perhaps even thousands of long-billed wading birds probing the mud for food. Curlew, redshank, wheeling flocks of dunlin and knot, godwits and turnstones all congregate alongside shelducks and geese to feast on the riches that the tide leaves behind. These are amongst the few truly natural habitats that remain. I find them profoundly fascinating and timeless places.

So it is a tragedy that climate change may destroy many of the best remaining examples.

The effect of rising sea levels on the fate of coastal habitats has been studied in some detail by scientists at the Natural Environment Research Council's Institute of Terrestrial Ecology (ITE) at Monk's Wood in Huntingdon. Their conclusions make disturbing reading.

Few people now doubt that sea level changes will occur, either through thermal expansion of sea water or through the melting of ice, where approximately half of the world's freshwater supplies are currently immobilized in solid form. The Climatic Research Unit at the University of East Anglia puts best estimates of the sea level rise within the range of 17 to 26 cm. by the year 2030. Some institutions are already making provision for the consequences of changing sea levels of this magnitude, and the National Rivers Authority is now planning all future sea defences for the East Anglian coast on the assumption that sea levels will rise by 5 mm. each year. This may be prudent, but it is still not enough to cope with the cautious estimates of sea level rise made by the Intergovernmental Panel on Climate Change (IPCC). They say that sea levels will rise by 650 mm. by the end of the next century, so even by these conservative estimates the sea walls built to the new specification would be inadequate by the year 2080.

The ITE report puts the cost of improving Britain's sea defences to resist a rise of one metre over the next century at a minimum of five billion pounds. There seems to be little doubt that this money will need to be spent, if only to preserve the productive, low-lying agricultural land of the Fens and the Humber Lowlands, which would otherwise be inundated.

This particular stretch of coast also happens to be the location of more than a third of all salt marshes in Britain. As a result, the cost of dealing with this particular consequence of climate amelioration would also need to be measured in terms of loss of wildlife habitat as well as in commitment of financial resources. Over 82 percent of our salt marshes are National Nature Reserves or Sites of Special Scientific Interest.

*

The inhabitants of the British Isles have been fighting a losing battle against the encroachment of the sea for centuries. Our coast is studded with groynes, breakwaters and sea walls which are designed to prevent saltwater invasion of valuable land and to control the erosive power of tides and currents. The only areas of coastline that are relatively immune to the incursions of the sea are rocky sea cliffs; elsewhere the waves constantly scour, sort, erode and redeposit sediments. Most of the coastline is in a state of restless turmoil. The soft clay cliffs of Holderness, north of Hull, disappear into the sea at a rate of over two metres per year. Within the last thousand years, thirty villages have plunged beneath the waves of the North Sea as this soft coastline has crumbled away.

One strategy for dealing with sea level rise, as far as wild habitats are concerned, would be to do the very minimum. If the sea were allowed to creep a little further inland, salt marshes and sand dunes would reform at a higher level. The maximum increment of sea level rise will probably be around a centimetre a year, and salt marshes can just about grow at this rate. But for many areas of the coast, such a *laissez faire* policy would be totally unacceptable. How many farmers would be prepared to donate a proportion of their land in the interests of perpetuating mud banks, sea lavender and scurvy grass?

Enhanced sea defences will also be necessary to preserve some prime wildlife habitats, like the low-lying Somerset Levels, which are threatened with flooding when sea levels rise. Conservationists have fought long and hard to preserve what remains of this marshy landscape of water meadows and drainage ditches that are a relic of a vanished agricultural system. Up until now the threat has come from attempts to drain the meadows; by the end of the next century the threat could easily be saltwater encroachment.

It seems certain that, wherever possible, sea defences will be strengthened. Where this occurs, the unique fauna and flora of the salt marsh will be slowly inundated and will disappear.

Amongst the plants that decline will be species like the strange, succulent glasswort, or marsh samphire. This vivid green species is constructed from swollen, jointed stems which can form colonies that stretch like an emerald green mirage over hundreds of square yards of shining, brown mud. It is edible and was once pickled. In Britain it is a major colonizer of bare mud which is regularly flooded by salt water, but in Saudi Arabia closely related species are cultivated in saline deserts, where plantations are irrigated with sea water and harvested for fodder and for their seeds, which contain a valuable oil. Glasswort may have a more secure future in the world's hot coastal deserts than on some parts of the British coastline.

And then there is sea aster, a scruffier version of the garden aster, but nonetheless delightful for its pretty blue and yellow flowers that extend over a salt marsh from the middle reaches to the dry land bordering fields. On a hot summer's day it often appears as a shimmering haze of pastel colour. Under its tall stems the smaller pink flowers of the creeping saltwort cling close to the mud.

Higher up the marsh the shrubby, grey-green sea purslane binds the mud and stabilizes the raised banks that are colonized by sea lavender and thrift. Thrift was familiar to generations of Britons who perhaps may never have seen it in the wild, since it used to figure on the back of the old brass threepenny pieces.

At the top of the marsh grows scurvy grass, not a grass at all but a member of the cabbage family, with masses of brilliant white flowers that appear in spring. Its glossy green rosettes of leaves have a high vitamin C content and are reputed to have come to the aid of sailors suffering from scurvy. Often, the delicate pink flowers of sea spurrey follow on from the white flowers of scurvy grass as the seasons progress.

Finally this vegetation grades into salt pasture, where less salt-tolerant meadow plants meet the salt marsh proper.

This succession of plants in the upper salt marsh seems to be destined to be squeezed out of existence at many points along the coast, sandwiched behind rising tides and sea defences that are impregnable to the salt marsh plants. There may no longer be a place for them in a warmer Britain. At present we have a hundred thousand hectares of this highly specialized habitat – an area of about a hundred and fifty square miles.

Aside from the loss of the plant communities themselves, several animal species that graze on the upper salt marsh vegetation, like brent and barnacle geese, will also need to move elsewhere in search of food. In the case of the geese this may mean that they will feed more intensively in farmers' fields, where they are already a cause of friction between conservationists and the agricultural community.

But worst affected of all bird species associated with salt marshes will be redshank. At the last count about 17,500 pairs nested in the middle and upper regions of salt marshes, representing about 60 percent of the total population. Also at risk is a small finch, the twite, which is already a rare species. This duller coloured version of the linnet spends the winter feeding on the seeds of salt marsh plants.

While the upper salt marsh seems destined gradually to slip below the waves, thousands of acres of mud flats, home to internationally important populations of sea birds, will also begin to disappear. Rising sea levels will reduce the area of available mud and also reduce the period of time that it is left uncovered by the receding tide, resulting in a shrinkage of feeding area for wading birds and ducks.

Mud flats are far less uniform than their featureless appearance might indicate. Subtle differences in the extent to which different areas dry out have a profound effect on the animal life below the surface, while the size of the mud, sand and silt particles determines which animals can survive. Some species, like cockles, have fairly catholic taste as far as the consistency of mud is concerned, but others are much more choosy. The muddiest areas hide ragworms, millions of tiny shrimp-like creatures called *Corophium*, and molluscs like the gaper and furrow shell, while sandy areas have a different group of shrimps, large numbers of the tiny nut shell and beds of tellins, filter-feeding molluscs that prefer a coarser substrate to burrow into.

If large-scale disruption of the salt marsh ecosystem sets in, the delicate equilibrium between erosion and deposition of particles will be shifted to an extent that will depend very much on local circumstances of wind and tide. Release of fine suspended particles into the sea will produce turbid water which will upset the delicate filter-feeding mechanisms of the inhabitants of the mud flats, while the disappearance of the upper salt marsh will lower the productivity of the whole system.

Whatever the details of this sea change in sea shore life, the net result will be not only a major loss of bird feeding grounds, but also a change in the menu for wading birds and wildfowl.

Many of Britain's estuaries are internationally important feeding grounds for waders and geese, and the species which would most feel the pinch as sea levels rose would include the eider duck and a long list of wading birds. As the mud flats shrink, the density of feeding birds will rise, and competition for food will become intense, leading to a steady reduction in their populations.

Another coastal habitat which will suffer badly from sea level change is the shingle beach, which often forms on the seaward side of salt marshes, near harbour mouths. If space permits they will erode in a landward direction, but when circumstances prevent this there will be some instances where they will slowly disappear. With them will go some intriguing plants, like the strange yellow-horned poppy, whose seed pods elongate like long, curved, green tusks after the flowers are pollinated, and sea kale, whose tough, glaucous leaves are crowned by a dense mass of brilliant white flowers.

Most of Britain's little terns, sometimes called sea butterflies because of their fluttering wing beats, maintain a fragile presence as a breeding species by nesting on these shingle banks.

While the outlook for salt marshes and shingle is less than optimistic, the future of some other important coastal habitats seems slightly more secure.

Sand dune systems should be relatively unaffected, although once again this will depend on whether there is space for them to migrate in a landward direction. Where this is possible they will reform; where their movements conflict with human interests the problems created will be much harder to resolve.

On the seaward side of land dunes there is one plant of the strand line which seems destined for a lean time as sea levels rise. Sea rocket is a pretty, pink-flowered annual in the cabbage family which has made the strand line almost its own. It shares it with a few other species of the fore-dunes that can survive amongst shifting sands and falling water tables, including the ugly prickly saltwort, the pretty sea milkwort and the uncommon sea holly. All of these species will be in the front line as sea levels rise. If the sand dunes move back, they too will be able to retreat higher up the beach. If not, they could be early casualties.

My own favourite stretch of sand dunes lies along the Northumberland coast, between the romantic ruins of Dunstanburgh Castle, perched on a craggy outcrop of black dolerite, and the tiny village of Low Newton-by-the-Sea. It is an exhilarating place, with wide, sweeping, sandy beaches backed by tall, undulating dunes that are higher than houses.

Hidden in their folds and hollows are pyramidal orchids, burnet rose, cowslips and bloody cranesbill. If there is one flower that can be said to epitomize this stretch of wild coastline, it is this magenta-flowered cranesbill that studs the fine, rabbit-grazed turf. Its colour is so vivid that even on the dullest of days, when the slatey-grey waters of the North Sea are whipped into white-crested breakers by icy north-easterlies, it seems almost luminescent.

If rising sea levels were to cause beach erosion and a landward retreat of these sand dunes, there would be a serious conflict of interests. The popular Embleton golf links and valuable farmland flank the landward side of the dunes. For the National Trust, which owns this sweep of coastline, there would be a tricky decision to be faced: balancing the interests of the golfers with the loss of prime wildlife habitat.

Similar conflicts of interest will be repeated all around the coast as the result of sea level changes, since golf courses have been laid out on the landward side of sand dunes in many places.

The first effects of this type of coastal erosion have already arrived. At Royal Troon in Scotland, sometime home of the British Open Golf Championship, combinations of high tides and gales are already eating away at the course. This may be distressing for golfers, but it may be even more unfortunate for the exquisite flora of lime-loving plants that the course is also famous for. Some experts expect to see the course disappear within the next forty years.

Many of the major coastal golf links are important sites for wildlife and half of the top fifty golf courses in Britain are Sites of Special Scientific Interest, acting as sanctuaries for rare plants like the dune gentian and lizard orchid and threatened animals like the natterjack toad and various reptiles. If these coastal courses fall victim to erosion, the security of these species is certain to be threatened.

However, it seems that there will be at least one positively beneficial outcome of sea level rises as far as sand dune systems are concerned. One of the dune habitats that hosts the greatest botanical riches is the dune slack, a wet area behind the foredunes which owes its botanical diversity to the fact that a layer of freshwater floats above the saltwater table. This particular plant community, of water plantain, bogbean with exotic fringed flowers that look as though they belong to a tropical flora, purple loosestrife, lady's smock, a rare wintergreen, marsh orchids and marsh helleborines, bog pimpernel and the scarce grass-of-Parnassus, depends on a constant supply of fresh water close to the surface. A rise in sea levels should help to ensure that this high water table is maintained, even if the summer climate becomes hotter and drier.

Before leaving sand dunes to their fate, I cannot resist mention of two unusual ferns that are frequently associated with this particular habitat. Neither looks like a conventional fern and both have fascinating connotations.

Adder's tongue fern consists of a small, flat-bladed leaf and a small spike of spore-producing tissue with an uncanny resemblance to a serpent's tongue. It is hardly surprising that the old alchemists, following the Doctrine of Signatures, firmly believed that this weird plant would cure snake bite. Should the Guinness Book of Records ever acquire a genetics section, this fern would merit a mention, for one adder's tongue fern species has more chromosomes than any other known organism – a staggering 1024. The commonest British species of adder's tongue, *Ophioglossum vulgatum*, has over 500. In contrast, we humans have a paltry 46.

Closely related to adder's tongue fern, but found in the shorter, drier turf at the rear of the dune system, is moonwort, *Botrychium lunularia*. This fern was another firm favourite of alchemists, who believed that it would turn quicksilver (mercury), or even base metals, into silver. Its single lobed leaf is supposed to resemble a mediaeval key and was reckoned to help in the opening of recalcitrant locks and chastity belts. The life of an alchemist must have been one long, frustrating round of total failures.

TWELVE

The Parable of the Water Flea

> And the fishers shall mourn
> And all they that cast angle into the brooks shall lament
> And they that spread nets upon the waters shall languish
>
> ISAIAH

There is no doubt that strenuous efforts will need to be made to resist the advances of the sea on to valuable land and that often, when the battles are lost, the results may be catastrophic. The population of the small North Wales town of Towyn have already experienced the effects of coastal flooding. In February 1990 an unfortunate conjunction of high wind and high tide led to massive breech of the sea wall, which flooded the town, making hundreds of local people homeless.

Our low-lying coasts will become more prone to this type of flooding as sea levels rise. The average change in sea level may only be of the order of half a metre, but when this is combined with high winds and swollen rivers discharging large volumes of water after heavy winter rain, the frequent outcome will be heavy flooding in estuarine areas. In 1953 a devastating North Sea surge occurred, flooding large areas of the East Coast. Even a modest rise in sea level will increase the frequency of such events.

When farmland is inundated with sea water it becomes useless for growing many crops for years afterwards, although some agricultural species are surprisingly tolerant of saline conditions. When Towyn residents returned to their homes they found that almost all plants in their gardens had been killed by the saline water, except for Brussels sprouts and winter cabbages, which appeared to be thriving.

Cabbages and their relatives, which come under the generic term of *Brassicas*, are direct descendants of the wild cabbage, a coastal plant which is still found in a few localities in Britain, like the cliffs near Staithes of the North Yorkshire coast. As a maritime species it is naturally more resistant to salt water and salt spray than most other cultivated plants.

Where farmland is temporarily flooded with seawater, the recovery process depends on the leaching out of salt by rainwater and a slow restructuring of the soil and re-establishment of the natural flora. In the interim, replacement of cultivated grasses with some suitable wild grasses, like salt marsh grass and red fescue, might well help to restore some of the grazing value of salt-damaged agricultural land.

The destructive effects of sea level changes will be most obvious around the fringes of our islands, but the influence of salt water will also be felt in ecosystems that lie some distance inland. All of our river systems ultimately discharge into the sea, and as sea level changes progress, saline water will penetrate further inland than ever before.

Storm surges have driven salt water up river estuaries and into freshwater ecosystems close to the coast on infrequent but regular occasions in the past, most recently in March 1988, when thousands of fish were killed by saltwater encroachment into the Norfolk Broads. These surges are the result of a number of factors, which include the coincidence of extreme high tides with high winds and barometric pressure changes, which together produce exceptionally high tides. Coastal defence measures often channel this sudden rise in sea level into low lying estuaries, resulting in severe flooding and an aftermath of salt damage.

The tidal range along the Norfolk coast is characteristically small, with only a two-metre rise between mean high and low water marks at spring tides. This means that a small rise in sea level along this coast represents a very significant rise in the mean high water level and makes the coast especially prone to storm surges. At other locations, like the Severn estuary, where the tidal range is over eleven metres, even a half-metre rise in sea levels is a relative drop in the ocean.

Saltwater encroachment into freshwater ecosystems like the Norfolk Broads is disastrous for wildlife. A limited range of estuarine animals, like eels and even sticklebacks, are tolerant of wide fluctuations in salinity, but for most true freshwater fish an influx of salt water is fatal. Mass fish kill has an immediate knock-on effect on animals higher up the food chain, like otters, kingfishers and herons. Many of the more spectacular organisms near the top of the freshwater food chain soon disappear.

Damage from tidal surges is transient and ecosystems eventually recover, but the subtler effects of prolonged saltwater penetration can be irreversible. A regular influx of salt water would change the bankside flora of the Broads, killing plants and adding to the erosion problem that already afflicts the intensively used parts of these popular waterways. The algal flora would quickly change, and with it the microscopic freshwater fauna, producing an ecosystem that would become estuarine rather than freshwater. Intertidal worms would replace dragonfly larvae. Seaweeds would replace waterweeds. The gradual decline of the characteristic Broadland vegetation would lead to a steady disappearance of many insect species, very few of which are associated with saltwater coastal ecosystems. The Norfolk Broads would quickly become brackish and infinitely duller, while the agricultural land that borders them would suffer from seepage of salt water and would also deteriorate.

This grim prognosis for one of Britain's best-loved waterways is just one facet of a more general threat to freshwater ecosystems in a changing climate. Even deep, landlocked lakes will be adversely affected.

A strange and perhaps unique fish swims in the deep, cold waters of Loch Ness and at the bottom of Lake Windermere in the Lake District. The arctic charr is a survivor from the last Ice Age and probably entered the inland lakes where it now lives as the climate warmed. Somehow, shoals were prevented from returning to the sea and they have been isolated ever since, swimming in the gloomy depths of the northern lakes for thousands of years.

Some of these populations have quickly evolved small but significant differences during their landlocked existence, to the point where charr from different lakes were once regarded as separate species.

Arctic charr in the British lakes are exclusively freshwater fish, but they do have close relatives that still swim in the arctic seas and which still ascend northern European rivers to spawn, in much the same way that the Lakeland and Loch Ness populations must have done before their retreat was cut off.

The survival of the lake-dwelling charr is linked to a peculiar physical property of deep bodies of water. In summer, well below the surface of the water, a horizontal discontinuity in water temperature develops, known as a thermocline. In this transitional zone, in the space of about a metre, water temperature falls suddenly by over 1°C, creating a boundary between warm water above and cold, well-oxygenated, nutrient-rich water below. The charr, which is a species with very specific temperature requirements, crosses this boundary occasionally, when it comes to the surface to feed, but by day it inhabits the cold depths of the lake.

One of the many consequences of climate change may be that the thermoclines of deep water lakes will be disrupted. This will be disastrous for animals that have become adapted to stratification in water temperatures, and particularly for those which inhabit the deeper, colder waters.

Several factors can contribute to increased mixing of the temperature layers, including wind speed across the surface of the water. Windier weather, coupled with higher temperatures of the surface layers of water, can force the levels of thermoclines down. Wind over the surface also creates turbulence, mixing layers and bringing nutrient-rich water to the surface, triggering dense algal growths. There is some evidence that such changes are already occurring.

A group of researchers at Canada's Department of Fisheries and Oceans at Winnipeg have been studying changes in lakes in north-western Ontario for the last twenty years. Over this period they have found that air and lake temperatures have increased by 2°C. Throughout the last two decades, lower than normal rainfall and enhanced evaporation rates have depressed water levels, so that concentrations of nutrients in the lakes have risen, encouraging algal growth around the shores. At the same time windier conditions have disturbed the surface waters and have deepened thermoclines, so that the summer habitats for fish that are choosy about their temperature requirements, that need the cold water below the thermoclines, have steadily shrunk. The water has also become clearer, aiding algal growth, because of the decreased volumes of water flowing into the lake; normally, fast-flowing streams would be stained with humus and would cause turbidity in the water.

The Canadian researchers believe that these changes in the Ontario lakes are an indication of what we can expect in deep bodies of fresh water when climate amelioration begins to exert wider effects. If they are, it will mean that arctic charr populations, which are already small, will be forced into a shrinking habitat and may well disappear from British lakes.

The British arctic charr population is already in deep trouble, due to acidification of its few remaining habitats. Crowding of the remaining fish into a smaller volume of water could help to administer the *coup de grâce*. Large numbers of arctic charr died in Loch Doon in September 1975, when abnormally dry weather caused water levels in the Loch to sink by about thirty metres. The charr were ready to spawn and already under stress, and the reduction in the size of their habitat and higher water temperatures led to an epidemic of disease in a crowded habitat. The pattern of climatic change that is predicted for future years could bring regular repeats of this mass mortality.

Lack of knowledge is a constant stumbling block when it comes to predicting the detailed biological effects of climate amelioration, but there are at least some detailed surveys of aquatic ecosystems that have been painstakingly maintained over many years, which allow some confident predictions of the kinds of changes we might expect.

One of the key animals in freshwater food chains in lakes and ponds is the water flea, *Daphnia*, familiar to generations of children who have enjoyed pond dipping. Water fleas eat algae and fish eat water fleas, so these tiny animals are a vital link in converting plant life in still waters into animal proteins.

The seasonal population growth and annual fluctuations in water flea numbers have been monitored in Esthwaite Water in Cumbria since the mid-1950s by biologists from the Institute of Freshwater Ecology at Windermere. Their findings tell an interesting tale.

The most important facts about the lake are these: it is stratified, with a summer thermocline that traps warm water above and cold water below; and, like many bodies of freshwater, it has become steadily more polluted with nitrogen and phosphorus from agricultural run-off and treated sewage.

The most important facts about the water fleas are that winter survivors begin to breed as the temperature of the surface waters rises in spring, and that their feeding depends on a good supply of minute, single-celled algae that are small enough for them to swallow.

This combination of water characteristics and water flea life style conspires to provide the animal with a limited period when it can breed and feed. The formation of the thermocline, raising surface water temperatures, controls the hatching rates of the eggs and the initial growth rate of the population, providing a 'temperature window' for the species that lasts about seven months but which peaks in July or August.

The peak season for the small edible algae, however, is much more restricted and they tend to decline as water temperature increases. As that happens they are replaced by blooms of large, inedible, blue-green algae which are far too large for *Daphnia* to filter out of the water. As a result the animals starve. The blue-green blooms have been fertilized in recent years by the nutrient pollution that has accumulated in the lake.

The upshot is that the peak water flea season, when the right combination of temperature and food is available, is relatively short. The late summer period in particular is usually dominated by inedible algae, especially if calm weather prevents wind mixing of the warm and cold water layers.

The influence of wind on water mixing seems to be the critical factor in determining the year-to-year variation in water flea abundance. Low animal numbers, brought about by blue-green algal blooms, coincide with years when the water stratification is stable, when calm weather maintains the temperature layers within the water column and prevents mixing. One theory attributes this stability to low wind speeds over the lake surface, which in turn are dependent on variations in pressure systems over the Atlantic.

So the abundance of the water flea in Cumbrian lakes seems to be related to weather conditions that might owe their origin to meteorological phenomena occurring hundreds, or even thousands, of miles away. And by identifying the chain of events that control the animals abundance, and perhaps that of the other inhabitants of the Lakes which feed upon it, the researchers are able to make tentative suggestions as to how climate change might affect the freshwater biology of Esthwaite.

One conclusion is that long periods of hot, still weather – the predicted summers of the future – which would stabilize the water temperature layers, will lead to blue-green algal blooms that will be bad for the water flea. Another is that the overall increase in temperature in the next century will lengthen the period when the water is warm enough for the animals to breed, stretching it to at least three-quarters of the year. These factors, together with the nutrient pollution of the lake, will determine the future pattern of water flea populations and, inevitably, those of other lake animals that feed on them. A seasonal change in the availability of water fleas would be almost certain to affect the growth and breeding performance of fish.

At least in the case of the water fleas of Esthwaite we have some idea of how climatic variables could influence the abundance of freshwater organisms. To reach this level of knowledge, the researchers have accumulated 12,000 water temperature records, counted 7000 water flea samples and analysed 6000 batches of plankton. Deciphering the interactions within natural ecosystems is a formidable task. With the rate at which climate change is now predicted to occur, it seems unlikely that other similar studies could be carried out fast enough to give us much inkling of how the new climate will affect habitats.

The unlocking of the secrets of the Esthwaite water fleas is one of a tiny group of research projects of its kind. When we try to cope with the biological changes that new weather patterns will bring we will rue the fact that, over the last few decades, those who have controlled the funding of biological research have assigned a low priority to attempts to understand the intricacies of the mechanisms of our natural environment. During a period when scientific research became increasingly focused on disciplines that offered the prospect of short-term economic benefit, it must have seemed that such investigations merely represented acquisition of knowledge for its own sake. What many of the decision makers in science still fail to comprehend is that the accumulation of knowledge and understanding of this kind is in our long-term interest. Without it, we have little hope of being able to continue to exploit technological advance in an environmentally sustainable way.

There are few types of vegetation in the British Isles that could truly be referred to as stable. The natural process of development in an ecosystem is for less competitive species to be replaced by plants that are more competitive, ultimately leading to an ecosystem where a complex assembly of plants exists in a form of equilibrium. Usually there is a trend towards increasing complexity in both the number of species in an ecosystem and their variety of growth forms, with both factors conferring stability. Tropical rain forests are the classic examples of complex, balanced ecosystems and in temperate countries their closest equivalents are deciduous woodlands.

It is the natural fate of all wetland habitats with shallow water and gently sloping banks to progress through a gradual sequence of ecological changes until they have become stable woodland. The rate at which this happens can be slowed by direct human interference or can fluctuate depending on climatic variables like rainfall and temperature. The pattern of climate change that is predicted for the southern half of Britain will lead to a rapid acceleration in the drying up of wetland habitats.

One recurring theme in any analysis of the effect of climate change on our wildlife is that in many instances it will be the ecosystems, plants and animals that are already threatened by our exploitation of their environment which will be worst affected. So it is with wetlands, which have rapidly disappeared through the pressures of changes in land use. Throughout the second half of the twentieth century, wetlands have undergone a precipitate decline, largely as a result of agricultural policies which have made it financially attractive to drain waterlogged lands, fens and ponds for the purposes of growing surplus food. Well over half of Britain's lowland fens have disappeared since the Second World War.

During the age of steam, ponds were dug in rural areas everywhere to slake the thirsts of traction engines, but since then the fate of these small standing bodies of water has been firstly to fall into disuse and then to silt up and become a last resting place for old mattresses and assorted domestic rubbish. The upshot is that newts, toads and frogs have declined catastrophically in many parts of Britain.

I find that one of the saddest aspects of the plight of our

countryside is that children can no longer be allowed to collect tadpoles, even if they can find them, because of the pressure on the frog population. It is but one of many examples of the way in which the destruction of wildlife habitat distances people from contact with nature.

Higher temperatures, with seasonal droughts, will make the future of ponds and other small bodies of standing water even more precarious. The gradual death of a pond during a sequence of summer droughts will follow a well-recognized pattern.

Plant species which colonize the margins of ponds are equipped with anatomical and biochemical adaptations which allow their roots to grow through oxygen-starved muds. These adaptations can take a variety of forms, ranging from hollow stems and air chambers, which transmit oxygen from shoots to roots, to biochemical mechanisms which detoxify products like lethal ethyl alcohol, which can accumulate in the tissues of roots growing in soils which are starved of oxygen.

This gives them a competitive edge over dry land plants lacking these adaptations. These cannot survive for long in the waterlogged soils of pond margins and are held at bay by the demanding physical conditions. But as the pond begins to dry up the balance is tipped firmly in their favour.

The partially submerged plants of the pond margin are not well adapted to drought, and when they are denied water for long periods they quickly die. Even before this stage is reached, dryland species can move in and colonize the thickening muds. Willows and alder are particularly adept at this and once established can grow several feet in a season. The further the water recedes, the more extensively they can colonize. As they become established, their roots draw on the remaining water supplies. During transient periods of rain, silt and plant debris build up around them, creating permanent dry land which extends towards the middle of the pond with each succeeding dry season. Slowly but inexorably succession proceeds, until the pond finally silts up and a characteristic type of woodland, known as fen carr, takes over.

The process I have described is a natural one, taking place gradually in all ponds, but sequences of seasonal droughts, which kill the established pondside vegetation, accelerate the process and prevent the usual degree of recovery which takes place in the wetter winter months. The potential speed of the change will have been obvious to anyone who visited any of the reservoirs which dried up during the dry summers of 1989 and 1990. Grasses and sedges colonized the bare mud almost as soon as it was exposed by the receding water and willows quickly spread into marginal areas of reed mace and burr reed.

There is another, particularly nasty, consequence of the drying up of ponds, ditches and streams which is easily overlooked: concentration of pollutants. Water contamination, with everything from sewage to agrochemical sprays, is a consistent factor which will worsen the specific effects of warmer summers in freshwater all over Britain, and this will be a particular problem with respect to the effects of drought on standing bodies of water. The effects of treated sewage and agricultural run-off into lakes, reservoirs and ponds are made worse by high temperatures, since these increase the evaporation rate and so concentrate the pollutants into a smaller volume of water. This chain of events will often culminate in algal blooms in shallow water, as nutrient concentrations increase. So climate change is almost certain to have very visible consequences on standing bodies of water of all kinds.

If seasonal droughts were to become a reliable feature of British summers, it would mean that many wetland ecosystems could only be maintained by intensive management practices, involving the use of sluices and the pumping of water. Without this continuous effort, whole communities of species – from fish and amphibians to pond weeds and dragonflies – would be under even greater threat than they are at present.

Some pond animals, like water fleas, are well equipped to survive seasonal drought, and although the adults may die, resistant eggs survive which can hatch when water returns.

But many of the most exciting and exotic aquatic animals are far more demanding.

One unusual animal that has already suffered badly in seasonal droughts is the Great raft spider. This is Britain's largest spider, which lives on the mossy margins of boggy pools, where it hunts down its prey. It was only discovered in 1956 and is confined to fenlands on the border between Norfolk and Suffolk. It has always led a precarious existence, and in 1977 the World Wildlife Fund supplied Suffolk Trust for Nature Conservation with sufficient funds to dig twenty-nine pits which should have maintained its habitat. But pumping of water from boreholes nearby during recent dry summers has steadily lowered the water table. In the summer of 1990 the pools had virtually no standing water and the spider population crashed dramatically. Hotter summers seem likely to continue this trend, so that the spider's days on British soil may be numbered.

The practice of continually pumping ground water in exceptionally dry summers has rapidly lowered water levels in many parts of the country, and by the autumn of 1990 those in some parts of East Anglia were as much as eight feet below normal. This inevitably leads to drying out of vulnerable areas of standing water and the loss of some of the most delicate and beautiful plant species of lowland mires and bogs, like insectivorous sundews.

Unless firm management practices are implemented, many of the best wetland sites for wildlife will disappear in lowland areas of southern Britain by the second half of the next century. Even with a substantial boost in the limited cash resources that are now available for nature conservation, it is hard to see how any but the most important sites will be retained. Even river systems are threatened: in the 1989 and 1990 droughts many of the minor rivers in calcareous lowland areas shrunk to a trickle or even dried up completely.

In northern Britain, where rainfall levels are expected to increase, the problem will be less acute, although patterns of rainfall are at least as important as total amount. Prolonged dry summers, with most of the rain falling in winter, a distinct trend that became apparent during the 1980s, may still lead to seasonal drying of smaller bodies of water which are important breeding sites for insects like midges, which in turn provide food for swifts, swallows and martins.

If summer rains should continue to be reliable in the north, wetlands there may expand. In this situation the best conservation policy may be to use available resources to develop and protect northern wetlands, perhaps introducing southern species, and to abandon attempts to preserve these habitats in the south, which may prove to be a fruitless and wasteful exercise.

THIRTEEN

Oceans of Ignorance

> South coast bathers have been warned of a plague of octopuses caused by mild winters. Staff of Dorset's Sealife Centre are warning the public not to touch the five-foot long creatures which have stinging suckers and beak-like mouths which can bite. Marine experts say the plague is concentrated on the French side of the Channel but could affect British beaches.
>
> *Guardian*, 27th May 1991

Speculation that climate change is the cause of a wide variety of changes in the flora and fauna, which cannot be immediately explained by any other means, is rife. The 1991 Spring Bank Holiday octopus alarm followed reports in late 1990 that there had been an octopus population explosion in the North Sea which threatened the livelihood of lobster fishermen. Apparently, almost every lobster pot pulled to the surface contained an octopus. The population of small lobsters, which would normally be thrown back to grow to a marketable size, was being decimated. The fishermen were convinced that the cause was the recent succession of mild winters which had allowed the octopus to breed in larger numbers.

Octopus plagues have been recorded after mild winters in the past, and it is well within the realms of possibility that the huge build-up in numbers was related to climatic factors. The species which invaded the North Sea is *Octopus vulgaris*, which has a natural distribution which extends into the Mediterranean and which usually reaches a northern limit in the English Channel. It may well be that its movement northwards is related to an influx of warmer water, which is said to have penetrated into the North Sea.

The arrival of this warmer water has also been implicated in the disastrous decline of sand eels, which in turn has precipitated a population crash of sea birds like puffins, which depend on sand eels for feeding their young. What the fisherman may have witnessed is a symptom of the climate amelioration that many scientists now anticipate: that the range of many animals will move northwards. This hypothesis has mainly been applied to terrestrial animals in the British Isles, but could be equally relevant to marine animals like the octopus, as sea temperatures rise and currents shift.

But there could, of course, be many other indirect causes for the octopus population explosion and the disappearance of sand eels. Evidence is the key to sound science, and scientific research is needed to establish the link between cause and effect.

Complex ecosystems are so large, and the interactions of all of their component organisms so complex, that establishing the true nature of links between species abundance and an all-pervading environmental influence like temperature will be an extraordinarily difficult task. Given the short timescale over which climate change is expected to occur, it seems most unlikely that much useful scientific evidence can be gathered before the changes take hold. The best that we will be able to do will probably be to monitor changes and explain them with hindsight, rather than accumulate evidence that will allow us to make any particularly useful predictions.

Scientific research tends to be concentrated on organisms which are convenient, accessible and amenable to experimental study and which may sometimes serve as a model for the wider effects of environmental change. Much of the detailed knowledge of biological organisms has been gleaned from a remarkably small sample of plants and animals which happen to be convenient to grow and maintain under laboratory conditions, or which are important for economic reasons.

The best understood animals are a fruit fly called *Drosophila*, and a minute roundworm called *Caenorhabditis*, both of which are amenable animals for laboratory experiment. Science has tended to become increasingly reductionist in its approach, so that we are close to understanding the molecular basis of the control of development in the fruit fly and the roundworm. Meanwhile, virtually nothing is known about the simple details of the life histories of the great majority of living organisms. The task of just naming the components of the living world is far from complete. One conservative estimate is that we have so far only described between a fifth and a tenth of the animal and plant species on Earth. It has been a trend of modern biological science that we have concentrated on learning more and more about fewer and fewer species. This may make for efficient research, but it has left us in a weak position with regard to being able to predict the effects of a gross change in physical environment on the vast majority of species.

The onset of climate change created by elevated levels of atmospheric carbon dioxide has demonstrated the depths of our ignorance about the natural world, especially in such inaccessible habitats as the oceans. We know little about the influences that are likely to affect the long-term future of their animal and plant populations. It is this shortage of systematically collected information that has made it so difficult for biologists to make projections about the effect of change. If nothing else, climate amelioration has emphasized the value of the scientific study of the natural history of our flora and fauna in allowing us to make rational decisions when planning for its future management.

One fine example of the kind of basic information that we will need comes from a study on hermit crabs which has been carried out by Ian Lancaster at Penwith Sixth Form College at Penzance in Cornwall. His methodical investigations have established, for the first time, the stimuli that trigger a male hermit crab's sex drive and his female counterpart's urge to lay eggs.

Using glass replica snail shells, Lancaster was able to watch the progress of sexual development of female crabs in aquaria and compile a detailed reproductive chronology. Hermit crabs only reproduce during the winter months, and females with eggs are only found between late November and early April, when sea temperatures fall within the range 6 to 8°C. As temperatures rise, they stop producing eggs. With a 43-day gestation period, females that develop an early batch of eggs can produce up to three broods per season.

By keeping the crabs in aquaria with varying temperatures and daylengths, Lancaster was able to show that it was temperature and not the hours of daylight that controlled female fertility. Sea temperature was not, however, a key factor in inducing mating behaviour. Males were stimulated to attempt to copulate by short winter daylengths in aquaria where summer water temperatures were maintained.

It is easy to see from simple but painstaking studies of this kind that changes in sea temperatures could have profound effects on the breeding biology of an animal like the hermit crab, whose egg producing activity is inhibited by elevated temperatures. Changes in winter sea temperatures could disrupt breeding cycles, at least until natural selection has had an opportunity to favour crabs with changed habits. Perhaps, in the case of the hermit crab, we may witness the disturbing sight of frustrated, randy males, stimulated by short days, pursuing confused females whose eggs have failed to develop because the water is too warm.

Changes in temperature in the large water mass of the open ocean can only be gradual, and, at least to begin with, slight. Temperature changes in the shallower waters of continental shelves could be larger and occur more quickly. But the most spectacular interactions between oceans and climate can have devastating effects.

The Peruvians and Ecuadorians are well aware of the cataclysmic potential of changes in winds and ocean currents. For much of the time easterly trade winds blow across the Pacific, from the South American coast towards Australia, driving currents of warm water towards Asia. The warm, wet winds bring monsoons to India and South East Asia.

At the same time, deep, cold water wells up along the west coast of South America, bringing a rich reserve of nutrients from the depths to the surface. The result is that for much of the time the coastal waters of Peru and Ecuador are amongst the most productive on the Earth's surface. The nutrients support the growth of plankton, which in turn feeds massive shoals of fish. Spectacular colonies of sea birds gorge on the marine life. It is one of the richest food chains known to science.

In the past, the nitrogen-rich droppings of the birds like the Guanay cormorant, which nested in tens of thousands on the Guano Islands, were used to fertilize European and American agriculture. These end-products of the rich upwellings of the Pacific deeps lured the crews of clipper ships to make the perilous journey under sail around Cape Horn to collect them.

The mineral-rich coastal waters once hosted a major anchovy industry. At its peak in 1971, the Peruvian fishermen removed twelve million tonnes of fish per year from their fishing grounds, much of it for export as fish meal to feed the cattle herds of rich nations.

But at regular intervals climate plays a cruel trick on these coasts. The easterly winds weaken, causing updrafts and severe storms, and the warm flow of surface waters slows. Sometimes, at Christmas, the wind and water currents reverse. Warm surface water flows towards the South American coast, colliding with the upwelling cold water and driving it back down into the deep ocean trenches.

The South Americans call this current *El Nino* – Christ Child – and it inevitably brings disaster. The loss of the nutrient-rich, cold water leads to death of plankton, fish and sea birds on a vast scale as the food chain snaps. Red tides of poisonous algae are fed by the decaying marine life. In 1972, *El Nino* triggered a decline in the acutely overfished anchovies and brought about the complete collapse of the industry. Animal feed prices rocketed in Europe, and Peruvian fishermen were ruined. Guanay cormorant populations shrunk, while the population of the Humboldt penguin declined by 65 percent after the 1982/83 *El Nino* deprived the species of food, making it an endangered species.

The 1982/83 *El Nino* was the most spectacular in recent history. The monsoons failed in India, droughts and forest fires occurred throughout Australasia, and torrential floods swept through Peru and Ecuador, raising some rivers to a thousand times their normal level. Climatic repercussions were felt over two-thirds of the globe, and severe effects registered as far afield as the West Coast of the United States, where heavy floods caused enormous damage. Fierce typhoons struck the western Pacific. Some two thousand deaths resulted and eight billion dollars' worth of damage to property ensued.

There are no regular current reversals on the scale of *El Nino* in the oceans around Britain, which is fortunate, because a radical change in the course of the warm North Atlantic Drift, colloquially known as the Gulf Stream, would probably bring a sub-Arctic climate to north-western Europe. But lesser changes in currents and sea temperatures can bring some unexpected visitors.

During the first week of August 1990, Britain and much of north-western Europe was firmly gripped by a blistering heat wave, for the second consecutive year. Speed regulations were imposed on inter-city trains when railway officials feared that the heat would distort tracks and cause derailments. A runway melted at Heathrow Airport. Woodlands caught fire near Cross Keys in Gwent. At London Zoo keepers hosed down sweltering gorillas and at Rotterdam Zoo the giant Amazonian waterlily, *Victoria amazonica*, flowered for the first time in eleven years. With temperatures reaching tropical levels, no doubt the massive plant felt at home.

It was, as newspapers are fond of proclaiming, a 'scorcher'. On Friday 3rd August the temperature in Cheltenham, Gloucester, reached 37.1°C (98.8°F), the highest since records began.

The period of fierce heat raised sea temperatures and brought some unusual marine creatures to our shores, perhaps aided by changes in winds and sea currents. Thresher sharks, a giant sunfish, red mullet and trigger fish were recorded in the English Channel, and a seven-foot-long Caribbean loggerhead turtle, well off course during its migration to the Mediterranean, put in an appearance off the coast of Dorset. It was the first loggerhead to reach Britain since 1938.

Even more remarkable was the invasion of Portuguese men-of-war. Coastguards warned bathers of the hazard from this, the world's most dangerous jellyfish, but not before a fifteen-year-old girl swimming in the sea at Goring in Sussex had been severely stung.

Although the Portuguese man-of-war, *Physalia physalia*, is commonly called a jellyfish, it is really a member of a bizarre group of closely-related animals called the Siphonophora. It is not one animal but thousands, each like a minute sea anemone.

Despite its appearance of having been designed by a committee, *Physalia* has a sinister, functional beauty. All that an unwary bather is likely to see is the gas-filled bladder, or float, like a foot-long Cornish pasty shot through with pink and blue iridescence. Below hang vivid blue contractile tentacles, sometimes reaching thirty feet in length, each composed of thousands of individual, microscopic animals which work together in perfect coordination. Some individuals are adapted for movement, some for reproduction, and some for feeding, forming a common stomach. And then there are those whose function is attack and defence.

It is the latter that make *Physalia* such a threat to swimmers, since their stinging cells are the most venomous of any known jellyfish. The excruciating pain is said to resemble that of a severe burn and was enough to cause the swimmer at Goring to collapse. The mere shock of being stung could easily be sufficient to cause a weak swimmer to drown. Accounts exist of sailors' arms becoming temporarily paralysed after hauling in ropes entangled in *Physalia* tentacles. Little wonder that in Britain, where venomous animals are rare, this drifting menace is regarded with fear on the rare occasions when it visits our coasts.

The Portuguese man-of-war has a worldwide distribution, and since it has no real means of swimming, it merely goes where winds and currents take it. Apart from tilting the float occasionally, to wet its surface, *Physalia* is a passive drifter. Those that arrive on our coasts are probably propelled from the equatorial Atlantic by south-westerlies associated with long spells of warm weather; during this century there have been invasions in 1912, 1919, 1921, 1934 and 1945, most of which were hot summers. The animals usually arrive in late summer, and during the large 1945 invasion hundreds appeared along the northern coats of Devon and Cornwall.

Unusually large or exotic marine animals and plants regularly reach our coasts during warm weather, but at present our waters are generally too cold for their liking. Exotic arrivals, like the sunfish, a disc-like fish which can weigh 1500 pounds and reach ten feet in diameter, often make national news, whilst less spectacular visitors, like tunny, pilchard and anchovy, are familiar to south-coast fishermen. In 1989 a fragile, short-snouted sea horse managed to make the perilous journey from the Mediterranean across the Bay of Biscay to Devon where it was caught amongst the eel grass beds near Drake's Island in Plymouth Sound.

Occasionally these exotic marine species linger, like the Mediterranean trigger fish that took up residence in the warm water outlet from Fawley power station in the Solent. And on rare but increasingly frequent occasions, they become established to the point where they integrate with or even dominate the native marine wildlife.

There seem to be plenty of suitable niches for introduced species in our marine flora and fauna, and we have a growing population of species from the Pacific which include the furry-clawed Chinese mitten crab, which is spreading from its original landfall amongst the mud banks of the lower reaches of the Thames. Other Japanese arrivals include a sea squirt (a member of a group of animals which have a very primitive backbone in their larval stages and are thus of great evolutionary interest), a tube worm and a sea spider.

Some of these may have arrived on the hulls of ships, or could perhaps have been introduced with transplanted Japanese oysters. Farming exotic species in shallow coastal waters is becoming lucrative, and inevitably some of these animals will escape. That certainly seems to be the case with the king-sized Japanese prawn, which commonly appears on the menu in Chinese restaurants; the first living specimens were caught by Brixham trawlers in 1990, and they had probably escaped from commercial farms on the French coast. An Australian barnacle, *Elminius modestus*, first appeared near Southampton in the 1940s, having probably arrived attached to a ship, and quickly established itself. It is still spreading rapidly along the Atlantic, English Channel and North Sea coasts. But this small crustacean is relatively benign compared with the threat posed by a particular plant invader from the Pacific.

Japweed, *Sargassum muticum*, is the most recent example of a marine alien that has become a pest. In the same way that our impoverished land flora leaves vacant niches for invading species, so there is room around our coastline for seaweeds from other oceans. *Sargassum* is exploiting one such niche to perfection.

Researchers at Portsmouth Polytechnic have been monitoring *Sargassum* since it first turned up on the shore at Bembridge in the Isle of Wight in the early 1970s. Its native country is Japan, and it was carried to the western coast of the United States and probably to Britain in consignments of Japanese oysters. Once established, it expanded into vacant ecological niches and flourished; here in Britain it may grow to a length of fifteen feet, six times longer than the plants in Japan.

Its rapid spread caused widespread concern and control measures were quickly applied. Its potential to choke harbours and compete with natural marine floras was an obvious threat, and when drifting masses of the weed converged on bathing beaches on the Devon and Dorset coast a concerted effort was made to eradicate it. Some of its prime haunts were coastal Sites of Special Scientific Interest, like the Bembridge Ledges, so chemical poisons were ruled out and the only method available was hand picking, a slow and labour-intensive process. At the same time a public awareness campaign was mounted in the media, asking for information of new sightings. Volunteers were recruited to help pull the plants up.

It was all to no avail, and by 1976 hand picking was abandoned. Research showed that the seaweed had no significant natural enemies here and the only remaining control means possible were cutting or sucking it from the seabed with a giant vacuum cleaner.

In a remarkably short space of time, *Sargassum* had become a fixture in our coastal flora. The pattern of its spread will be familiar to anyone who has studied the progress of the more aggressive terrestrial alien plants. Although its establishment and spread here has little directly to do with changing climate, it dramatically illustrates that our islands contain many vacant niches that are suitable for alien species, so that a change in temperature could favour the aliens rather than established natives. As our seas become warmer we might expect to encounter more alien marine organisms taking up residence around the coastline. Giant kelp, imported from California and deliberately planted around the coast of France as an industrial source of alginates for use in the food industry, may soon join Japweed as a pest along the English coastline.

The fishermen whose incomes were threatened by the octopus invasion in the summer of 1990 had already become accustomed to the disastrous consequences of a combination of rising sea temperatures and marine pollution. In the late spring of 1990 their foe was not the octopus, but a microscopic plant which poisoned their catches.

The culprit was an alga, which produced a spectacular 'bloom'. Algal blooms, which include the notorious red tides that have poisoned hundreds of people in seafood-eating nations like Japan, are caused by rapid multiplication of these plants and occur when the water temperature rises and nutrients are abundant. In the case of the North Sea, the nutrients probably came from agricultural run-off and sewage discharge, producing a nutritious broth for algal proliferation.

The algae appear as a red film on the tide line, or a reddish dust on the water surface. One of the commoner species is *Gonyaulax*, which produces toxins which accumulate in mussels that filter them out of the water. Dining on such poisoned shellfish can paralyse the nervous system or even kill.

At the height of the algal bloom, a Ministry of Agriculture ban was placed on selling not only filter-feeding shellfish like mussels, but also prawns, lobsters and crabs. Tests showed the presence of the toxins in these crustaceans, no doubt because they had been feeding on mussel beds. The poison was passing up the food chain. At the same time holidaymakers were warned not to touch the red scum on the tideline.

Selling any of the affected animals for human consumption was temporarily forbidden along the whole stretch of coast between the Humber and Montrose. Three thousand fishermen and a thousand inshore boats were unable to work, bringing many close to commercial ruin.

Algal blooms have been a feature of years when mild winters and hot summers have warmed polluted coastal waters. If this weather pattern were to become commonplace, coastal fisheries may well be permanently lost. Paradoxically, tourists who might be drawn to the British coast by the improved summer climate would be ill-advised to venture into the sea when the water was at its warmest. Such a prospect emphasizes the need for a concerted effort to clean up the waters of the continental shelf.

Exceptionally warm summers have also favoured another nasty organism, botulism bacterium, *Clostridium botulinum*, which has had a devastating effect on some sea bird populations.

Type C botulism is a well-established disease of waterfowl in California, and in exceptionally hot years it has also wiped out large numbers of coastal birds at other locations, including the Camargue. It first turned up in Britain, in the Firth of Forth, during the exceptionally hot summer of 1976, when hundreds of dead gulls were picked up along beaches. The final death toll was probably over four thousand birds.

The birds became infected through feeding in contaminated pools in municipal rubbish tips. The bacterium produces a toxin which is exceptionally poisonous. Three or four blowfly maggots, feeding on the decomposing corpse of a bird killed by botulism, can accumulate enough of the toxin to kill an adult pheasant. So birds like herring gulls, which scavenge maggot-infested carcasses, might well pick up and ultimately transmit the disease via blowflies.

The toxin affects nerve endings and produces a progressive weakness. The birds stagger, fall over and die within about twenty-four hours. Some acquire immunity, but the disease has flared up again in recent hot summers.

Botulism from decomposing food virtually wiped out the herring gull population on Steep Holm in the Bristol Channel during the blistering summers of 1989 and 1990. Gulls passed on the poison to chicks, which died in the nest. Breeding pairs fell from ten thousand in 1975 to about a hundred in 1990. If we are to experience regular hot summers in the future, we can expect this pestilence to remain a problem for sea birds and waterfowl. It will be one of the more unpleasant consequences of hot summers.

The death of sea birds from the virulent botulism toxin caused concern amongst ornithologists and environmental health officers, but it only merited short paragraphs tucked away in the corners of the pages of a few daily newspapers. It passed more or less unnoticed, in stark contrast to the massive outcry from the public when dying seals, infected with a mystery disease, began to appear on our television screens.

On rare occasions a wildlife issue such as this arises which receives press publicity and arouses intense feeling at a national level. The seal plague that visited our coasts in 1988 spawned pictures of helpless, dying seals which struck a chord with the general public in a way that destruction of marshland by motorways could never hope to. They generated such intense feeling that the government was forced to increase its meagre support for measures to understand and combat the disease. Nothing could have focused attention more acutely on the chronic pollution of the North Sea. It was as if public opinion had awoken from a coma.

Most of our population have probably never seen a seal in the wild, but television pictures transmitted into living-rooms all over the country made people feel very uncomfortable. The large, dark and intensely feminine eyes of a dying seal have a great emotional appeal, of the kind that a diseased cod could never arouse. If those almost-human eyes were registering such pain, could we soon be at risk? Perhaps there was a realization that the health of our wildlife might act as a litmus test for the health of the environment, and for us in turn. The fate of a fellow mammal hung in the balance.

Although a seal may look like a heap of wobbling blubber from a distance, those eyes reach into the soul in a way that pollution-sensitive molluscs, destroyed by the toxic anti-fouling paints from ships' hulls, or remote sea bird colonies, starving through the disappearance of sand eels, could not. Such anthropomorphic considerations play a large part in conservation policy when it comes to public pressure and government response.

The dying seals did more for the cause of wildlife conservation than any number of earnest environmentalists could achieve. And for as long as the cause of their illness remained a mystery the finger was firmly pointed at marine pollution.

Eventually a scientific consensus arose that the seal plague was due to the action of a naturally-occurring and previously unrecorded distemper virus, related to the virus which causes human measles. No conclusive direct evidence could be presented for the widely-held suspicion that marine pollution had weakened the seals and made them more susceptible, something of a relief for a government which was still fighting a rearguard action to continue using the North Sea as an extension of the national sewerage system.

But some scientists are convinced that pollution was a factor. The pollution in question was atmospheric carbon dioxide enrichment, a product of our quest for economic growth.

David Lavigne at the University of Guelph has re-examined the mass mortalities that occurred and believes that climate is indirectly but strongly linked to seal deaths. He has pointed out that it was probably no coincidence that almost all recorded seal plagues have occurred during periods of unusually hot weather.

During exceptionally warm weather, seals all around the coasts of Britain, Holland and Denmark haul themselves out of the sea to sunbathe on sand bars. Lavigne and his colleagues have pointed out that a dense colony of sunbathing seals on a hot summer's day creates perfect conditions for the rapid spread of a disease epidemic. A sand bar on the tideline of an intensely polluted shallow sea, inhabited by several hundred seals, lolling shoulder-to-shoulder as temperatures soar, is as favourable a breeding ground for seal diseases as Victorian slums of the 1880s were for typhus, cholera and the other epidemics that strike human populations.

So the dying seals may well have been victims of a combination of climate change and marine pollution, a natural disease and their own innate behaviour patterns. Our contribution has been to allow the environment and conservation issues to remain so low on the political agenda for so long. If the recent summers are an example of what we can expect in future, and if seal population densities recover to their pre-1988 levels, we can expect to see those wide-eyed helpless stares on our TV screens again.

The episode of the seals encapsulates a fundamental problem we have with nature conservation, namely tinkering with the symptoms of the overall malaise, rather than developing a strategy to tackle its root cause.

Seal numbers rose throughout the 1980s because of an emotional wave of public revulsion at annual culls, said to be necessary to conserve commercial fish stocks. The seals that escaped slaughter in this way died of disease at least partly because of our inability to preserve their environment in a healthy state. Species conservation without consideration for the wider environment is a step on the road to ecological ruin.

Perhaps the latest public emotional response, this time reacting to the horror of the seal plague rather than to the dubious necessity to carry out culls, will take us a step further towards realizing the fundamental defect in public attitudes to wildlife conservation, which are orientated towards species rather than habitats. Saving seals with inoculations and seal sanctuaries salves the conscience of a society that measures the value of its natural heritage mainly in monetary terms.

Dealing with the real problem, of curbing the destruction of wildlife habitat, requires a major change in our relationship with our environment and the sacrifice of ideals based on the exploitation of natural resources for maximum economic benefit. It will be economically painful, at national and personal levels.

FOURTEEN

Fading Genes

> Suddenly there burst forth a general cry of 'Here they come'. The noise which they made, though yet distant, reminded me of a hard gale at sea, passing through the rigging of a close-reefed vessel. As the birds arrived and passed over me I felt a current of air that surprised me. Thousands were soon knocked down by the pole-men; the birds continued to pour in; the fires were lighted and a most magnificent as well as wonderful and almost terrifying sight presented itself. The pigeons arriving by thousands alighted everywhere, one above another, until solid masses as large as hogs heads were formed on the branches all round. Here and there the perches gave way with a crash, and falling on the ground destroyed hundreds of the birds beneath, forcing down the dense groups with which every stick was loaded.
>
> JOHN JAMES AUDUBON, describing the annual arrival of migrating passenger pigeons in Kentucky. The last of the species died in Cincinnati Zoo in 1914.

Perhaps it needs the eye of an outsider to draw attention to the real plight of our flora and fauna. I think that it was Paul Hogan, Australian star of the movie *Crocodile Dundee*, who remarked that he had never visited a country where there were so many wildlife conservation organizations and so little wildlife left to conserve.

We have a plethora of conservation bodies, some of which are large and influential. English Nature, the successor to the Nature Conservancy Council, is the government-appointed organization that provides the strategy for managing our wildlife heritage, and is staffed by professional biologists. Its funding and effectiveness are a direct reflection of national priorities with respect to wildlife.

Otherwise, the task of nature conservation is fragmented amongst a wide spectrum of charities. Many of these have permanent professional staff but rely heavily on charitable donations and voluntary help. Many of the most important are small: even the Royal Society for Nature Conservation, which considers itself the leading organization in wildlife protection, has only about two hundred thousand members; about four thousand for each of its constituent County Wildlife Trusts.

Wildlife conservation has always operated in a fragmented, piecemeal way under a siege mentality, fighting the tide of development that has swamped the countryside by rescuing the best remaining areas for wildlife. This is often achieved with the help of charitable appeals which fund the purchase and maintenance of nature reserves. The successes of the conservationists who have managed to protect so much of our flora and fauna in the face of overwhelming odds should never be underestimated, particularly since they have been forced to work with pitifully small financial resources.

But, because conserving wildlife has not traditionally been a primary consideration in planning for the development and use of our landscape, the most important elements of our flora and fauna are now confined to small nature reserves and Sites of Special Scientific Interest which are marooned between vast tracts of urban development and intensive agriculture. We have arrived at this situation because wildlife conservation in Britain is essentially a rescue operation, reacting to threats as they arise.

The whole concept of maintaining nature reserves is a tacit admission of our inability to live in harmony with wildlife. By setting aside small areas of carefully selected habitat, where what remains of our flora and fauna can survive relatively unmolested, we can create a convenient classification of countryside into small areas that should be left alone and large areas that can be managed for maximum economic gain, where wild animals and plants merit little or no consideration.

Now that the climate is changing, we are about to realize one of the major drawbacks of applying this strategy. For by isolating populations of plants and animals and banishing them into reserves, and by disrupting the continuity of populations, we have rendered them especially vulnerable to change. A graphic example of how fragmentation of animal populations places them at risk comes from the case of the Kaiserlautern frogs.

Complex road systems in industrialized countries link cities, towns and villages by a network of highways which divide up the countryside into a mosaic of islands set in a sea of concrete and asphalt. The transport infrastructure may not be a serious barrier to movement for animals that can fly, but for those that hop, slither or run it may confine their movements like the bars of a cage.

One result is that animals that once had a continuous distribution are subdivided into smaller populations which are more or less isolated from their neighbours. Population geneticists have long suspected that this fragmentation of animal populations may have serious long-term consequences for their genetic constitution. A study in Germany on the common frog has demonstrated that this is indeed the case.

Researchers at Mainz University have used enzymes from embryos in frog spawn to demonstrate that frog populations in the lowlands around the city of Kaiserlautern are not only genetically isolated, but are also inbreeding, leading to the erosion of genetic variability in local populations.

Enzymes are chemical substances with a dual function. On the one hand, they can act like biochemical chain saws, cutting up large molecules like proteins, fats and sugars into smaller molecules; on the other, they can act like biochemical arc welders, joining small molecules into larger, more complex ones. Isoenzymes are minutely varying forms of an enzyme, which all have similar catalytic properties, but which vary slightly in size or electric charge.

This means that isoenzymes migrate at different rates in

an electrical field. Biochemists can exploit this characteristic by loading them on to a porous gel and passing an electric current through it, so that the isoenzymes are sieved into a series of bands according to their size and electric charge. Then they can be made visible by staining with a variety of coloured reagents. The end-result is a kind of biochemical fingerprint which is also an index of genetic variability. The more variations in isoenzyme bands between individuals, the more variable the population.

This is a convenient method for measuring changes in genetic structure of populations, and the scientists at Mainz applied the technique, known as electrophoresis, to demonstrate that 88 percent of the isoenzyme systems in their frog populations existed in more than one form, each of which was under genetic control. The frog population had an intrinsically high level of genetic variability for its enzyme systems.

But their studies also showed that subpopulations differed dramatically in the frequency of the different forms of isoenzymes that they contained, suggesting that isolation by roads was leading to independent evolution in each population. Subpopulations were losing genetic variability, at varying rates depending on the number of individuals that they contained.

Worse still, it was clear that some of the genetic variability, manifested in different isoenzyme complements, had disappeared altogether, especially in the smaller subpopulations. This was almost certainly the result of inbreeding amongst a few surviving individuals in populations that were isolated from other frog colonies by roads.

Inbreeding in small populations has the nasty effect of exposing lethal genetic defects which would otherwise remain hidden in the population and only surface on rare occasions. Once these begin to emerge with a high frequency, the effect on a small population can be catastrophic. Not surprisingly, many organisms have complex mating patterns and genetic systems which tend to maximize opportunities for cross-breeding between unrelated individuals and thus avoid the evolutionary risks of incestuous mating.

Studies on toads have shown that low traffic densities, of less than thirty cars per hour, are sufficient to kill all animals that attempt to migrate across roads to spawning sites. But the Kaiserlautern biochemical study demonstrates that there is a more insidious, indirect threat to amphibians, in addition to being squashed by traffic. The fragmentation of their populations is leading to irreversible, local extinction of genes. In the study area, the main motorways are only thirty years old – about twelve frog generations – so the rate of this genetic erosion has been rapid.

So what is so worrying about the loss of a few genes? The answer is that almost all organisms carry a reservoir of genetic variation, which is their insurance policy in a changing world. And the world is changing faster now than it has in the last ten thousand years.

In the case of the Kaiserlautern frogs, the researchers point to one simple solution to the problem that they have identified. If habitats on either side of roads were linked with ditches and tunnels, the continuity of the frog populations and their genes would be assured. In the case of many less common and less mobile plants and animals, which are widely scattered on nature reserves, the problem is less easily solved.

Recent studies, like those published by the Institute of Terrestrial Ecology, suggest that even the modest changes in mean temperatures and rainfall predicted by the Intergovernmental Panel on Climate Change will have a profound effect on the distribution of individual elements of our flora and fauna. While few species will become nationally extinct, many will undergo regional decline. Species that are confined to south-facing slopes subject to prolonged summer droughts are amongst the highest risk groups; we might well see the demise of several rare plants from some of their best-known localities in the southern half of Britain.

Meanwhile, in the northern counties an amelioration of the climate will change the competitive balance between plants that are better adapted to cooler, wetter climates and those more aggressive species that can flourish in drier summers and warmer winters. In the main, rare species with finicky habitat preferences will decline, whilst common species, which are often abundant and widespread because of their undemanding nature, will increase.

These changes in status, with some common species becoming commoner and some rare species becoming exceptionally rare, have implications that stretch far beyond the immediate impact that they may have on the pattern of wildlife in Britain. They will also change the genetic structure of populations. In the long term, this may well be one of the most profound and permanent legacies of climate change.

Almost all populations of individuals are genetically variable. This is a banal enough observation as far as the human race is concerned, where our daily interaction with fellow citizens depends on the fact that we can recognize each other by our external features, but it is easy to forget that individuals within populations of trees, bees, birds and fleas, to name but a few, all carry an equally large store of heritable differences.

These difference are most apparent in domesticated animals, where extremes of genetic variability are deliberately selected. It is sometimes hard to believe that breeds of dog as different as Yorkshire terriers and Rottweilers are members of the same species, but it is so.

The extraordinary range of external appearance of the domesticated dog is testimony to the hidden range of genetic variation in the species, which can be ruthlessly selected by breeders who are looking for bizarre characteristics. Wolves all tend to look very similar because of equally ruthless selective forces operating in the natural environment, where weird aberrations of the kind that dog owners prize would be at a distinct disadvantage.

Natural selection tends to work towards producing individuals that are uniformly well adapted to their environment, while maintaining a hidden reservoir of genetic variation that can surface in times of change.

This variability in visible characteristics is merely the tip of a genetic iceberg. Many differences are invisible, manifested only as small changes in the molecular forms of enzymes, like the isoenzymes of the Kaiserlautern frogs or, ultimately, in the genetic material itself. Such molecular differences, which are exceedingly common in man because of our social taboos, which prevent inbreeding within families, form the basis of molecular fingerprinting techniques used in criminological investigations. Rapists and murderers are convicted by the products of their own unique set of genes.

Many genes contained in human chromosomes can exist in more than one form and the potential number of permutations of these genes is, in practical terms, infinite. No wonder life on Earth is so diverse.

This vast store of genetic variation is the key to continuity of life and is the raw material for evolutionary advance. When environments change, those individuals that by chance contain the most fortuitous combination of genes are subject to natural selection and are favoured. They compete better in a crowded habitat, breed more successfully and so propagate their sample of the species' genetic variability at the expense of weaker individuals, which die out and carry their genetic variability with them into extinction. The frequency of genes in populations varies continuously in space and time, depending on how useful they are to the organism in any particular set of circumstance at any particular moment.

Such is the kernel of Charles Darwin's theory of evolution. It is a process which has proceeded continuously since the first appearance of the living, self-replicating molecules in the organic soup that existed on Earth three thousand million years ago. By a vast game of chance it has generated us.

Cataclysmic climatic changes in past geological epochs have led to mass extinctions of groups of organisms and their genes. The disappearance of the dinosaurs is the best-known example, but there have been many others. During the Hirnanthian glaciation 439 million years ago in the Ordovician Period, the

Earth cooled dramatically. Patrick Brenchley, at the University of Liverpool, argues that this disruption of the Earth's climatic zones triggered three waves of mass extinctions. In the first phase, as the temperate seas became too cold to support some forms of life, trilobites, many starfish and some planktonic organisms disappeared. In the second phase the formation of ice led to a catastrophic lowering of sea levels in tropical regions, leaving tropical reefs and brachiopods (the primitive shellfish of the period) high and dry. In the third phase the global climate warmed again, sea levels rose, and the coral reefs, which had reformed after the ice age had lowered water levels, were drowned again in deep meltwater. All this happened in the space of half a million years, although the final, warming event may have been compressed into the space of perhaps ten thousand years.

Even within the worst case scenarios predicted by climatologists, nothing remotely as catastrophic as this is predicted in the current round of man-made climate change, although the picture for the spread of arid areas in subtropical regions and for the flooding of shorelines might bear some comparison. What these climatic changes in geological time do demonstrate is the capacity of climate to change the course of evolution.

The loss of the trilobites' genetic stock wiped out a complete line of evolution. Had it continued, a lineage of animals might have evolved which would have taxed the imaginations of present day science fiction writers. This is certainly so in the case of the extinct animals of the Burgess Shale rocks, a group of outrageously improbable organisms whose demise and rediscovery has been entertainingly described by Stephen Jay Gould in his book *Wonderful Life*.

These creatures, suddenly extinguished for reasons which are unknown, have the appearance of having been constructed from spare parts of today's animals. As Gould points out, if their disastrous demise had not occurred, forcing evolution into the pathway which has led to the present range of organisms, then the living world might look quite different now. If the genetic clock had not been reset by their extinction and if their genetic stock had persisted and evolved continuously to the present day, our ridiculously anthropocentric perception of the pinnacle of evolution might not have been a human being in its current recognizable form.

These prehistoric extinctions are relevant to our present predicament inasmuch as they illustrate the potent interaction between the physical environment and evolution. In the current round of environmental change, our more immediate concern is not with mass extinctions, but with a drastic narrowing of the genetic variability within living organisms and the consequent loss of genetic diversity.

Picture the following scenario, concerning a hypothetical butterfly species, whose caterpillars feed on an uncommon plant species which grows on south-facing slopes of the Sussex Downs. Agricultural intensification has slowly whittled away the butterfly population, until only one secure population remains, in a nature reserve.

Several years of prolonged summer drought have gradually reduced the caterpillar food plant population, and the butterfly colony is dwindling with it. Clearly, the loss of the food plant will soon lead to extinction of the butterfly, unless something is done. Conservationists watch with growing concern. But then, as the last remnants of the population seem doomed, an individual arises with a complement of genes that mean that its caterpillar can feed on an alternative, common, drought-tolerant food plant. At the eleventh hour, genetic variation and natural selection have saved the day. Evolution has triumphed.

Such a scenario, with a happy ending, is not very realistic. As a population's size decreases, its store of genetic variability shrinks rapidly, making the likelihood of last-minute salvation increasingly remote. Every cycle of inbreeding is another, tighter turn in a downward spiral of genetic erosion that drastically narrows the range of evolutionary options open to future generations, for whom the genes may be vital when new environmental conditions prevail. So a more probable outcome would be extinction of the butterfly.

It is precisely this kind of cycle of genetic decline that some biologists believe contributed to the disappearance of the large blue, only the fourth British butterfly to become extinct in historical times. The last surviving colony of this mysterious butterfly, whose caterpillars are tended by ant foster parents, petered out in 1979. Slowly, colony by colony, the large blue was forced into retreat from south-facing chalk and limestone escarpments in southern England until one small stronghold remained on Dartmoor. As the population size shrank, so its genetic uniformity rose, diminishing its ability to cope with environmental change. It finally disappeared after the severe droughts of 1975 and 1976, when lack of water killed the patches of wild thyme that the caterpillars fed on and also forced the ant guardians to eat the last large blue caterpillars that they had carried into their nests.

Armed with this stark prospect of genetic erosion, conservationists who hope to tackle the threat from climate amelioration are faced with some serious dilemmas. How long should they spend charting the decline of vulnerable rare species before they try to intervene? Should they intervene at all? If so, how should scarce resources be allocated? Who will decide on the priorities? Should pretty butterflies have priority over ugly slugs?

As the food plant and butterfly species decline, should the food plant and its dependent caterpillars be transported to a new site, in the hope that both can flourish? Or should nature be allowed to take its course, in the hope that natural selection will sieve out a butterfly with the right genetic equipment to allow it to survive? How much of our present-day flora and fauna should we endeavour to rescue? The range and complexity of these questions is bewildering.

Since the current round of environmental change is man-made, many would argue that we have a moral responsibility to intervene and try to minimize the effects of change on the living world. Having meddled once, we should meddle again, to try to rectify the situation.

Others might draw comfort from the long-term perspective of geological time, point to the comparatively massive changes triggered by far greater natural environmental perturbations of the distant past, and propose that we should now let nature take its course.

The latter non-interventionist argument is certain to win the day. It will appeal to governments who baulk at the enormous cost of the piecemeal, speculative nature conservation that will be needed in order to try to counteract changes which are predicted but cannot be guaranteed. It also makes some practical sense. In spite of the fact that Britain's flora and fauna has been more intensively studied than that of any other part of the Earth's surface, we still know too little about the precise environmental requirements of most of our plants and animals and their complex interactions to be able successfully to manage them in any precise and predictable way.

In reality we probably have little option but to try to minimize change in the environment by reducing its cause, atmospheric pollution. We must then accept the changes in our natural wildlife that may occur. We have started an experiment which we can slow down, but are unable to stop.

Ultimately our only option may be to create environmental zoos and botanic gardens, which will be museums of past habitats and their occupants, while the ecosystem at large is left to its own evolutionary devices.

But if we do embark on this *laissez-faire* course of action, we need to be aware of the long-term consequences. Climatic change will initially reduce the size of wild populations of organisms drastically, by selecting for survival those fortuitous individuals that have been handed a winning ticket in the genetic lottery. This subset of the whole population will then increase, but unique genes of those that were not selected will be lost.

Population geneticists have studied this phenomenon for decades and are well aware of the drastic reduction in genetic variability that can occur in small, closely inbreeding populations that arise in this way. Such populations face particular hazards if the environment continues to change, since their small subsample of genetic variability may no longer contain enough genetic potential to ensure their continued survival.

Exactly the same logic applies to populations of plants and animals that are kept in gardens and zoos. These too are small subsamples of populations, through economic necessity. No zoo could afford to maintain populations of polar bears that would be large enough to contain all the genetic variability present in the species.

One further difficulty in conserving genetic variability concerns timing. How long should we wait before taking action? Given the large amount of genetic variation available in large, interbreeding populations, how will we know when we have lost a few billion potential gene combinations? Such numbers are beyond comprehension, but in the competitive world of interactions between organisms and their environment they do have significance.

An analogy that would be apt in the current economic climate would be to compare such a change with a drastic fall in the value of a company on the Stock Exchange, wiping out a third of its assets. Just as companies become more vulnerable by the loss of financial capital, so living organisms become weakened when their genetic capital is eroded.

With hindsight, it seems that many of these conservation dilemmas might well have been avoided if we had viewed our flora and fauna as an essential part of the countryside and made more effort to maintain continuous populations of plants and animals throughout their natural range, rather than concentrating on maintaining small, isolated populations in nature reserves.

It is virtually certain that when the climate changes, we will find that our nature reserves are in the wrong places. Concentrating all your best wildlife resources in a single location when the climate is changing fast is a recipe for disaster. Better to employ the strategy proposed by the German researchers who have studied the Kaiserlautern frogs, offering living organisms the opportunity to migrate, uniting small populations into a single continuous breeding unit. Unfortunately, if many of our native animals were to migrate out of nature reserves, they would find themselves in totally unsuitable wildlife deserts of sterile farmland, suburbia or industrial development. We have left our wildlife with nowhere to go.

Things might have been different had we made a greater effort to live alongside nature, rather than pushing it into convenient corners. Had we begun to create wildlife corridors through developed areas, linking prime wildlife sites, half a century ago, rather than creating the wildlife ghettos that are our nature reserves, population shrinkages caused by climate change in part of the species' range might not have threatened species themselves. And those that could migrate would migrate, either as individuals or by passing their genes through the population.

But of course such speculation is fruitless. We plan the use of our landscape in terms of human life spans and not in evolutionary timescales, and the rapidity of impending environmental change has taken everyone by surprise. We have to live with the consequences.

In reality, we are unlikely to be able to formulate any rational basis for intervention in the genetic erosion of wild plant and animal populations until a catastrophic decline in levels of genetic variability has already occurred. Until rarity threatens species' existence, managing their genes is unlikely to be worthwhile in the short term. Because the logistical problems are so severe, we will always operate in reactive rather than proactive mode, as far as wildlife conservation is concerned.

When plant and animal species are of sufficient economic or aesthetic importance, there are technological fixes available which will allow us to bank valuable genes. For many, but by no means all, plant species, low-temperature storage of dry seeds offers a means of conserving a massive amount of genetic variability in a relatively small space. Some seeds have a very short life span and will not withstand freezing. For these plants, small pieces of tissue can be grown in culture and frozen in liquid nitrogen at minus 196°C. Such techniques of suspended animation have already been used for food crops like strawberries and cassava. Plant pollen grains and the eggs and semen of animals can also be stored in this way.

There are even more radical solutions available. One is to extract the DNA, the material that contains the genetic code in the cells of all organisms, cut it with enzymes into small sections and insert these into a bacterium called *Escherichia coli*, culled from the human gut. Countless colonies of bacteria, each with a cargo of genes from the organism, can be held in storerooms on agar culture plates. The complete genetic identity of a bird, butterfly or tree could one day be stored in this way, on nutrient jelly in a small plastic dish in a refrigerator.

Practical perhaps, but small halos of microorganisms set in jelly would make a poor substitute for the exquisite beauty of a swallowtail butterfly. It would be rather like scraping the paint that Van Gogh used to paint his sunflowers from the canvas, separating the powder into colours and pushing it back into tubes. We would still have the raw materials, but we could never recreate the authentic masterpiece.

Let's hope that it never comes to that.

FIFTEEN

The British Baked Bean

> There is no need that the plant kingdom cannot supply. There is no challenge to which it cannot rise. All that is required is a measure of human ingenuity, intelligence and the political will to care for it.
>
> ANTHONY HUXLEY, *Green Inheritance*

Until quite recently, any landscape painter producing pictures of the British countryside in summer would have relied heavily on the cool greens in the paintbox, with occasional recourse to splashes of yellow ochre for ripening grain crops. The colours of our landscape are dominated by the greens of lush grass and hedgerows and woodlands.

Our cool, wet climate is admirably suited to growing cereals and grasses; our average wheat yields per hectare are roughly twice those of the United States. The arable farming industry's tendency to specialize in mass production of this limited range of crops is largely responsible for the form and colour of our lowland landscape. But now change is in the air.

Over the last forty years British farming has responded to the call for higher production and cheaper food with a degree of efficiency that is probably unprecedented in any area of the national economy. This has been achieved partly by drastically reorganizing the landscape in the most fertile parts of the country: the patchwork of small fields has been replaced with an open landscape, which is in effect a massive food factory. The public, understandably, are uneasy.

We have all benefited from more efficient food production:

we all now spend a lower proportion of income on food than ever before; but there is growing disquiet over the environmental impact of the worst excesses of current agricultural practices, ranging from nitrate pollution of groundwater to wholesale destruction of wildlife habitat. There is a ground swell of opinion that says that we should change the way we farm the land.

Since the Second World War the phenomenal success of the appliance of science to farming practice and crop breeding, coupled with inept economic and political management of agricultural production, has led to catastrophic overproduction of most agricultural commodities. Belatedly, the European Community is working towards a better-planned, balanced agricultural economy, necessitating a review of the price support system for the major overproduced arable crops.

The resulting financial squeeze in the arable farming industry has encouraged a search for alternative crops, sometimes with industrial rather than food uses. The rapid development of plant biotechnology has also opened up new vistas for plant breeders, creating opportunities for developing 'designer crops', fitted to any purpose and adapted to any climate that will support economic plant growth.

The net result has been that British arable farming was already beginning to change before climate amelioration became a serious issue. The new thinking on agricultural crops must now be placed in the context of significant climatic shifts that are likely to occur within one farming generation. Take these factors together and it is clear that the pattern of arable crop production, and the colour of the landscape, is likely to change more in the next century than it has in the last millennium.

If a Norman farmer could walk through today's agricultural landscape in Lincolnshire, he would notice many remarkable changes that have occurred since his farm was recorded in the Domesday Book. He would certainly wonder where all the trees had gone, wonder why so few people worked on the land and marvel at the size of the fields and at the absence of weeds.

And he would probably be appalled at the disappearance of game and wildlife.

One familiar aspect of the agricultural landscape that would make him feel at home would be the crops, most of which were common in his time. Wheat, barley, oats, peas and beans would all have been grown on a Norman farm. Or on a Neolithic farmstead, for that matter. Their yields have increased and the methods for cultivation have developed beyond recognition, but the plants are still recognizably similar.

The two crops that would be novelties to a time-travelling Norman would be oilseed rape and sugar beet. Oilseed rape is an agricultural phenomenon of the late twentieth century. In the space of twenty years it has transformed our countryside. In many arable areas the predominant colour of the May agricultural landscape has changed from cool green to vivid yellow. Landscape artists must now venture into the unfamiliar territory at the yellow end of their palettes.

Tradition has it that the Dutch engineers who drained the Fens introduced oilseed rape into Britain, using the oil to lubricate their drainage pumps. Until the early 1970s it remained a relatively rare sight, confined to small areas in East Anglia and the East Midlands. Now the sea of yellow flowers extends well beyond the Scottish border.

The cause of this stunning change in the rural colour scheme stems from the European Community decision that Europe should become self-sufficient in edible oils to satisfy the huge demand from the food industry. Besides direct use in cooking oils and margarine, rapeseed oil finds its way into cakes, biscuits, ice cream and a wide variety of processed foods.

Plant breeders responded to the EC directive by producing new varieties that lacked the erucic acid that had previously confined the use of rapeseed to industrial applications. Aided by EC price support subsidies, the acreage of the crop has soared in the current decade. Britain is now the second largest producer, after France, and generates the highest yields in the whole European Community.

The other crop that would seem unfamiliar to our Norman time-traveller wandering through the British agricultural landscape would be sugar beet. He might just recognize the leafy tops, which had been used by the Romans as a vegetable, but he would be amazed by the swollen roots.

Sugar beet is an example of that rare agricultural phenomenon, a new crop. Until comparatively recently sugar was a rare and expensive commodity, and until the middle of the eighteenth century almost all sugar came from tropical sugar cane imported by sea. In 1747 a Prussian called Markgraf discovered that the roots of *Beta vulgaris*, the cultivated version of wild sea beet which still grows around northern European coasts, contained sugar in their sap. Sugar beet development proceeded apace and the first extraction plant was opened in Silesia in 1801. The Prussians maintained an early monopoly on the crop, until it was introduced into France at the end of the eighteenth century. It was in France that sugar beet became a major agricultural crop, as a direct consequence of the Napoleonic War with England.

The British navy successfully blockaded French sugar cane imports from the West Indies. In 1811 Napoleon issued legislation which not only made sugar beet cultivation a legal requirement, but also required the setting up of research centres for the study of the crop. Breeders quickly succeeded in selecting larger roots with higher sugar content. As the importance of the crop rapidly increased, imports of cane sugar from the colonies became less significant, setting in motion a chain of events which led to the decline of the sugar cane industry in the tropics. The successful European sugar industry is a direct legacy of Napoleonic agricultural policy. The effects of the Emperor's legislation are still being felt in cane plantations in Africa, the Caribbean and the Far East.

If there is a lesson to be learned from the history of sugar beet and oilseed rape, it is that major new crops are developed as a result of political exigencies. Left to its own devices, the agricultural industry would probably prefer to stick to what it currently does best, and grow grasses. But now that the climate is changing, physically and politically, the impetus has been created to develop alternative crops.

The success of oilseed rape has stimulated great interest in oil crops, most of which produce their best oil yields in warmer climates. One that has already begun to make an impact here is linseed. This produces a slow-drying oil which has a wide range of industrial uses, including the manufacture of linoleum, and its delicate slatey-blue flowers already grace several hundred hectares in the north of England. It is a visually attractive crop which most people would welcome in the countryside. The flowers open in the morning and begin to fade by midday, later being replaced by small, spherical capsules full of glossy black, lens-shaped seeds.

Linseed has been here before and was formally an important fibre crop, grown to produce flax. In the sixteenth century, when Britain's climate was warmer, cannabis, or Indian hemp, was also a significant British fibre crop. Studies of subfossil pollen preserved in peat deposits and lake sediments have shown that cannabis was probably extensively grown in eastern England, and records of legislation passed in 1533 reveal that Henry VIII insisted that each farmer cultivating more than sixty acres should sow a quarter of an acre of hemp or flax to provide the rigging for his navy.

Kevin Edwards, from Birmingham University, and Graeme Whittington, from the University of St. Andrews, have identified Scottish sites where cannabis was cultivated and have shown that the period of maximum hemp cultivation in Scotland coincides with a period of climatic warming and greater rainfall, perhaps of the kind that we might be about to enter. Cannabis cultivation seems to have died out in Britain as the climate cooled, when it may have been replaced by flax. If the narcotic content of the cannabis crop could be eliminated by breeding techniques, it might once again make a useful fibre crop at some time in the future.

Flax, now known as linseed, is cultivated for its oil content rather than its fibre, although improved retting techniques, to isolate the fibres from the stems, might also encourage a revival in flax fibre production.

Sunflower, on the other hand, is grown entirely as an oil crop, although the air-filled pith in the stems was temporarily used to fill life-jackets for sailors during the last war. The great, plate-like flower heads are a common sight in France, and throughout the 1980s plant breeders improved the climatic tolerance of the crop so that it spread steadily northwards, creeping closer to the Channel coast. We can probably expect it to make the Channel crossing and appear as a crop in southern Britain early in the next century. Elsewhere in Europe it has already become a popular crop; in 1990 the acreage of sunflowers in Germany increased by 63 percent and yields rose by 20 percent, helped by a long, dry period during harvesting.

The most significant consequences of climate amelioration, as far as British agricultural crops are concerned, is that the changes in temperature will lengthen the growing season and redistribute the rainfall. This will allow crops like sunflower, which would normally flower and mature too late in our climate to produce a commercial crop, to progress steadily northwards.

The speed of this northward migration is hard to predict, but according to Dr Martin Parry of the Atmospheric Impacts Research Group at Birmingham University, we can expect the areas suitable for cultivating maize as a grain crop to migrate northwards at the rate of about ten kilometres a year as the growing season lengthens. Currently, grain maize ripens reliably only in the extreme south-eastern corner of England, and is grown elsewhere in the south only as a silage crop. A 1.5°C increase in average temperature would mean that maize would ripen as far north as south Yorkshire and Lancashire; a 3°C rise would see maize crops ripening in the central Highlands of Scotland.

This lengthening of the growing season opens the door to several new legume crops. There is likely to be a steady revival in interest in these dual-purpose plants, whose roots contain bacteria which convert atmospheric nitrogen into nitrate fertilizer, as the trend towards organic agricultural methods gains impetus. This built-in fertilizer factory means that relatively little nitrogenous fertilizer is needed for growing legumes and that following crops benefit from the nitrogen they leave behind in the soil.

At present our best legume crop is probably the field or faba bean, the small seeded, agricultural version of the broad bean. It has had a chequered history. Field beans were traditionally grown on most farms when horses provided motive power, as home-grown animal feed; the medium-sized seeds that were preferred for this purpose were known as horse beans. The crop declined as the internal combustion engine replaced the horse and mixed farming gradually disappeared, although it has shown several brief signs of revival.

In the early 1970s a disastrous anchovy harvest along the west coast of South America, partly a consequence of overfishing and partly caused by the warm *El Nino* current disrupting marine life in these coastal waters, led to a drastic shortage of fish meal for animal feed, which briefly boosted the fortunes of the field bean crop. More recently, the acute problem of bovine spongiform encephalopathy, better known as mad cow disease, which has been caused by feeding infected offal to cattle as a protein source, has led to a resurgence in the demand for vegetable protein in animal foods and renewed interest in the crop.

Field bean is a visually attractive plant, with richly scented flowers that are particularly fragrant in the early morning. As a student I spent three years carrying out research on the crop, and the smell of a glasshouse full of flowering beans still lingers in my memory. It is the most fragrant of our agricultural crops and legend has it that travellers who slept beside field bean crops would be driven mad by their overpowering fragrance. Bumblebees love it.

Agriculturally, the crop has several defects which limit its yield and reliability, but one of its most important shortcomings is that although it has a high protein content, its oil content is markedly inferior to soybean, which is therefore the preferred source of seed protein in animal feeds. Now that the climate is changing, faba beans should become more popular in more northerly counties of England where soil moisture will not be a problem and where increased sunshine will speed the ripening of a crop that is normally slow to mature.

In the sunny south we should soon see a revival of interest in soybean. Its cultivation has been attempted here before on a pilot scale, but until now our growing season has been too short, with too few warm days to ensure reliable flowering and grain set. Attempts were made in the 1970s to market a cultivar called 'Fiskby V' to amateur gardeners that failed miserably, but with long hot summers the outlook for this crop, both as a legume grain crop and as a source of vegetable oil, is far more promising.

The same is also true for another legume, navy bean. This is a small, white-seeded version of French bean and is the key ingredient of baked beans in tomato sauce. Navy beans are currently imported from the United States, where they are grown on a large scale in Michigan, but there have been attempts to test the plant in our climate. In the 1970s Dr Richard Hardwick evaluated a range of cultivars at the National Vegetable Research Station at Wellesbourne in Warwickshire and also carried out breeding work aimed at reducing the incidence of halo blight, a bacterial infection of pods which afflicts the crop in this country. A warmer climate and the need to diversify cropping patterns should lead to the advent of the British baked bean as an important crop.

Many of these new crops are species that would almost grow here now during exceptional summers when conditions are particularly favourable. Other minor crops currently under investigation are longer-term prospects.

Agricultural lupins, which have fewer flowers and much larger pods than garden lupins, are grown for animal feed, and several cultivars are already planted on the Continent. One variety of Russian origin has shown some promise in the British agricultural environment.

There is also a growing range of minor crops for specialist uses. These include the attractive evening primrose, which has large lemon yellow flowers and tiny seeds which contain the medicinally valuable gamma linolenic acid, used in treating heart conditions. The blue-flowered comfrey, an extremely popular crop with bees, has larger seeds which also contain the same valuable compound.

Most of these minor crops are grown to satisfy a small, specialist market and are unlikely to have a major impact on the landscape, but they do represent a return to the kind of agricultural diversity that was a feature of farming half a century ago.

In the 1950s my grandmother worked on a very traditional farm which grew statice as an additional source of income. This is the cultivated version of sea lavender and it grew well on the flinty soils of my part of Sussex. It must have been grown in the area for generations, and was originally produced to satisfy the florists' trade, most of the crop being sent to London by rail. In late summer the fields were covered with a mist of slatey-blue flower heads on wiry stems, which looked wonderful against a background of ripening corn. Helping her cut the flowers as they began to open fully, tying the bundles and carting them off to dry in a large, black-timbered barn was something I looked forward to in the school holidays.

Minor crops like this, which add much to the countryside, may once again become locally important as sources of agricultural revenue as farms diversify. They would do much to temper public criticism of modern agricultural monocultures. It is the sheer monotony of the late-twentieth-century agricultural landscape that is the most obvious target for public criticism.

One of the most intriguing crops to be considered for British agriculture in recent years is quinoa, an ancient Andean crop which was once cultivated by the Incas. It is related to the weed called fat hen and also to the old pot herb Good King Henry. Quinoa is a broad-leaved grain crop, with heads of small seeds that are made into a porridge or added to bread in its native South America. One of its attractions here is the quality of its protein, which contains an exceptionally fine balance of amino acids that would satisfy the needs of vegetarians. Scientists at Cambridge University have been working on the development of the crop for Britain for several years.

In the longer term, a warmer climate presents the prospect of growing some far more exotic crops. In 1990, trial crops of sorghum, a sub-tropical grain crop that is widely cultivated in Africa and the southern United States, were tested in East Anglia. The intention was not to harvest sorghum grain, since the crop will not flower in our long days, whatever the temperature, but to see how much biomass it would produce.

Somewhat belatedly, the European Community is showing great interest in alternative, renewable and clean forms of energy that could supplement and reduce dependence on fossil fuels. The aim with the sorghum was to see just how much growth it could make that could then be converted into energy by a variety of processes, including fermentation to produce alcohol. The researchers chose a good year to begin their experiments, and in the hot summer of 1990 the crop did remarkably well, with plants reaching a height of over six feet. Sorghum can produce as much as nine tonnes per hectare of sugar, which can be converted into ethanol as a petrol substitute. Similar, so-called gasohol fuels are already widely used in developing countries, like Brazil and the Philippines. If the trials, which are also being conducted in Ireland, Belgium, France, Germany, Italy, Spain, Portugal and Greece, are successful, Britain could be growing energy crops on land which is set aside from cereal production.

The concept of growing plant biomass to exploit for energy production is a well-established one which is undergoing a revival, although its future is tightly linked with the price of fossil fuels and of oil in particular. At present and in the immediate future it seems unlikely to be a financially viable system, but past experience with shock rises in Middle East oil prices could quickly rewrite the cost/benefit equation. It would seem logical that we should invest in the development of these alternative energy sources at a time when fossil fuel prices are low, so that we can exploit them when fuel prices rise in the future.

Another crop that is being investigated for the same purpose is coppiced willow, a fast-growing tree which can be cut on a regular cycle. Dr John Porter at Long Ashton Research Station has assembled a collection of over one hundred species and varieties, and at least ten of these have considerable potential for biomass production. They are easy to propagate from cuttings and do well on land that is too wet for arable crops.

Long Ashton has a history of research into willow, and shortly after the Second World War the success of its scientists in this field prompted irate questions in the House of Commons. Half a century ago England had over nine thousand acres of osier beds, with the main production area centred on Sedgemoor. Many of the osiers had been coppiced and maintained for generations to supply cane for baskets and other utensils in the days before plastic displaced natural products. By the early fifties the area of osiers had shrunk to two thousand acres and the home-grown supply of cane had been replaced by imports from Argentina. Ironically, the Argentine osier industry owed its success to vigorous willow hybrids which had originally been bred at Long Ashton and exported to South America.

Today, only a tiny area of land is under osier production, to supply craft workers, but if the crop undergoes a resurgence as a source of energy we could see willows become part of the agricultural landscape once more. Within the pattern of rainfall that is predicted for the future, it would seem that the centre of production for the crop would need to move northwards.

Harvesting takes place in winter, during a period when farm activity is low and at the time of year when heat and power are most in demand. Trials show that up to twelve tonnes of wood per hectare can be harvested on a three- to five-year cutting cycle, producing wood with a high calorific value. The willow 'fuel chips' yield between half and a third of the caloric value of coal and oil respectively. One aspect of this form of energy production that is a strong selling point at a time when atmospheric carbon dioxide levels are rising is that growing willows absorb as much carbon dioxide as they release during combustion, so the net release of carbon dioxide into the atmosphere is virtually zero.

The prospect of beds of willows reappearing in the countryside is an attractive one. Willows were formally grown in 'withy' beds, regularly-spaced plantings of coppiced stools that regenerated whippy stems. Withy beds have all but disappeared, which is a pity, because they were valuable habitats for wildlife, with boggy ground that had a characteristic flora of marsh marigold, lady's smock and celandine and harboured bird species like water rail, marsh warblers and snipe. Diversification into crops of this nature will help to restore the variety of habitats in the landscape, and can only make a positive contribution to the future of wildlife.

SIXTEEN

The Technological Fix

> Science is a first-rate piece of furniture for a man's upper chamber, if he has common sense on the ground floor.
> OLIVER WENDELL HOLMES

Perhaps we should find reassurance in the rapid yellowing of the British countryside under a blanket of oilseed rape. It does, after all, demonstrate the capacity of the plant breeding industry to respond to change. Over the next half century agricultural scientists will need all the skills that they can muster.

Bearing in mind that the transformation of oilseed rape from a minor to a major crop in Britain has been achieved with conventional plant-breeding methods in less than fifteen years, this represents an encouraging sign of agriculture's ability to respond to change relatively quickly. Even more encouraging is the progress being made in plant genetic engineering, which will revolutionize the way in which plant life is exploited.

By the middle of the next century, when the climatic changes will be sufficiently severe to force a major reorganization of arable farming, new methods in plant biotechnology will have cut the time taken to breed a new crop variety by at least a half. The day of the designer plant is almost here.

Traditional techniques for plant breeding are labour intensive and painfully slow. They are as much a triumph of logistics as of science. To produce a new wheat variety the breeder selects the best available cultivar and crosses this with other plants which have desirable traits, such as disease resistance or tolerance to environmental stress. From the offspring of these crosses he then spends over a decade attempting to select plants which were as good as the original variety and which have the new gene added. This inevitably involves screening vast numbers of plants.

Breeding a new crop variety with current methods is comparable to having to completely dismantle a car every time it needs a minor service, mixing all the parts with those of another vehicle of a different manufacture and then attempting to put it back together again without a manual. The whole process for a cereal like wheat or barley takes about twelve years. In 1990 plant breeders made cross-pollinations that will produce the crop's varieties for the early years of the twenty-first century.

Climatic change poses a major problem for the plant-breeding industry. Breeding a crop for disease resistance or improved response to nitrogenous fertilizers is comparatively straightforward. Each potential variety can be exposed to an artificial fungal epidemic or to a particular range of fertilizer treatments and its relative performance measured.

Breeding a crop for an unknown climate that may exist in ten or twenty year's time is a different problem altogether. The worst-case scenarios for climate change predict average temperatures rising at a rate that will make it extremely difficult for conventional plant breeding to keep pace. The industry will be aiming at a constantly moving target.

Only two experimental approaches are available that will allow the breeder to subject the plants of today to the climate of tomorrow. One is to use a phytotron. This is a building filled with growth chambers, each simulating an agricultural climate. Cost alone would rule out their use for large-scale testing.

The other method is to breed crops for British agriculture at trial sites overseas, in a climate like that of southern France, which might approximate to south-eastern England's climate in the middle decades of the twenty-first century. This may be a satisfactory approach as far as simulating temperature and rainfall patterns are concerned, but it has one serious flaw. To breed plants in a climate which approximates to that predicted for Britain, we would need to go south towards the equator, where days are shorter.

Plants are highly sensitive to daylength. While higher temperatures can accelerate flowering, it is often daylength that triggers the switch from vegetative to reproductive growth and ultimately controls the yield of a crop. If a crop stops producing leaves too early and starts to flower too soon, the yield will be small, because there will be too few leaves to supply the developing seeds. If it takes too long to switch from leafy growth to flowering, the seeds will be produced too late in the season.

This means that a new cereal variety developed in southern France may well be drought-tolerant, but its overall agricultural performance might be quite unacceptable in the longer days of the English Midlands. Without testing the new varieties in representative locations in Britain it will be well-nigh impossible to assess their true performance.

The history of the cultivated potato gives a graphic illustration of the importance of daylength to crop plants. Potatoes are indigenous to the Andes, and the first cultivars which Spanish conquistadors introduced into Europe sometime around 1570 were adapted to the short daylengths of central South America. In Europe the long daylengths meant that they produced tubers far too late in the season to be of any agricultural value; some varieties produced no tubers at all. It took over two hundred years for a genetic variant to arise, either by deliberate breeding or perhaps by chance, that was capable of producing worthwhile tubers in the long days of a northern European summer.

This sensitivity of crops to even quite small differences in daylength means that selecting varieties for future agricultural conditions in northern Europe by planting them in the south is unlikely to be very successful. In addition to the problem of daylength, local differences in soil type and microclimate can also have major effects on crop performance, further complicating the issue.

But even if these approaches could be used, breeders are limited by the fact that they cannot precisely identify the target they are aiming at. The vague nature of current computer models is such that no one can predict with any accuracy what the agricultural climate of the next decade will be, let alone that of the mid-twenty-first century. Producing a range of suitable varieties that would adequately cater for all the nebulous possibilities of future climates would be exorbitantly expensive and wasteful. Such a level of speculative investment would not be a commercially attractive proposition for Britain's plant-breeding companies.

Plant genetic engineering may offer a way out of this dilemma. Genetic engineering of plants has the potential to cut the time involved in producing a new crop by about half and the labour by a far larger factor. This is already a major commercial consideration, at a time when the overproduction of cereals in developed nations means that seed companies must look to cutting production costs, as well as increasing sales, for profits.

Genetic engineering offers the opportunity to take the best available cereal variety and insert a single gene, so that a new, improved variety could be available to farmers in six years or less. It is a method for correcting defects without sacrificing all of the other valuable attributes of a crop variety which have been carefully nurtured over endless cycles of conventional breeding. This slick process of defect correction avoids the complete mixing and subsequent unscrambling of the genetic contributions of two parents. The whole process is analogous to upgrading a piece of electronic equipment by inserting a new microchip. The saving in time and money is enormous. The commercial world has not been slow to recognize this.

The early teething troubles of plant genetic engineering techniques were mainly concerned with performing this technological fix on plants in the grass family, which includes the majority of the important crop plants. The technique originally revolved around the use of a bacterium, *Agrobacterium tumefaciens*, as a Trojan Horse.

Genes were inserted in the bacterium, then the bacterium was allowed to infect plant cells. There it transferred its cargo of genes into the host. Unfortunately *Agrobacterium* could not be induced to infect grasses, nor could many cereals be regenerated from cells into whole plants, an essential step in the process.

More recently a technique has been evolved where microscopic tungsten bullets, coated with the DNA that encodes the desirable genes, are fired into host cells. Improbable as it may sound, the 'magic bullet technique' works remarkably well. Tissue culture techniques for grasses have also improved rapidly and the way is now open for adjusting the genes of our existing major crop plants for a constantly shifting climate.

The technology for plant genetic manipulation has developed at a faster rate than anyone could have dreamed at the beginning of the 1980s. The major problems in exploitation are now centred on public acceptance of the end-product, on ensuring that the commercial companies that develop the technology can exploit it commercially, and last, but by no means least, on identifying the genes that need to be transferred into crops to ensure their viability in the new environments of the next century.

The presentation of plant genetic engineering to the public has been badly bungled.

In 1983, when three laboratories simultaneously succeeded in transferring foreign genes into a tobacco plant, using the microorganism *Agrobacterium tumefaciens* as a go-between, it signalled the birth of a new technology. The scientists had broken down barriers to the exchange of genetic material between groups of plants which had become isolated over vast periods of evolutionary time.

The technique offered geneticists the opportunity to isolate single genes and switch them between plant species at will. In effect, there were no longer any barriers between species. Genes from beans could be inserted into cereals, genes from bacteria transferred into tomatoes. Even genes from animals could be inserted into plants. Science could perform tricks that were beyond the scope of evolution.

The commercial potential for the new technology seemed endless. In 1983 genetically engineered microorganisms and animal cells were already fairly commonplace, but exploiting the technique in bacteria or animals was fraught with difficulties. Releasing genetically engineered microorganisms into the environment would always be a sensitive issue, too reminiscent of opening Pandora's box for easy commercial exploitation. Genetic engineering of animals raised moral issues that would be bound to slow down or even block public acceptance.

In contrast, plants seemed safe. Even worst-case scenarios, with genetically modified plants becoming rampant weeds, did not seem to be too frightening a prospect. After an initial period of hesitancy, large transnational corporations began to pour money into the new technology and there was a scramble amongst agrochemical giants to buy plant-breeding companies.

The major problem was to identify and select the genes that should be transferred. In the early 1980s the environmental debate was slowly getting underway and the future of agriculture still seemed to lie in continued intensive use of agrochemicals. So the agrochemical companies, who largely carried the development costs of the new technology, looked for applications that would allow them to sell more of their products, a strategy which had proved a commercial winner in the earlier Green Revolution.

Herbicide resistant crops, which could be blanketed with weedkillers without suffering damage, seemed like a good idea. Such crops even presented the possibility of selling a follow-up herbicide, to kill 'volunteer' crop plants that germinated from residual crop seeds, which would act as a bridge to carry diseases over from one season to the next.

By the time these crops were ready for field testing the environmental debate had moved on apace. The agrochemicals industry had become the focus of intense criticism, and plants that were designed to grow under even higher levels of chemical treatments were nothing less than a public relations 'own goal' for the new technology.

At the same time, animal genetic engineering had taken a disturbing turn, with stories emerging of pigs crippled with arthritis after being given growth hormone genes. A new genetically-engineered hormone that was injected into cows to increase milk yields, called bovine somatotropic hormone (BST), also seemed to be a totally pointless piece of technology when Europe was awash with milk.

All this happened as public confidence in science was on the wane. There was a growing suspicion that the higher ideals of science were being overwhelmed by the tempting financial rewards from the novel technology. Even some scientists began to question the wisdom of using such methods to shore up creaking agricultural systems.

The American biotechnologist Martha Crouch pointed out that plant scientists are caught in a moral dilemma. So far they had managed to develop agricultural technologies which masked the symptoms of the greater environmental problem which threatens the global ecosystem: human population growth. By hiding the symptoms they had deflected attention away from the root causes of environmental destruction.

If plant genetic engineering was successful in offsetting the worst effects of climate change, might it not also weaken our resolve to tackle the cause of the problem?

Much was also made of the benefits of genetic engineering for the Developing World, but no one was sure who would pay for it. In order to make profits from their products, the plant-breeding companies would need to have tight control over their product via a legally enforceable patenting system. This raised all sorts of problems.

Firstly, enforcing patents might well prove to be impossible in the Developing World. Many Third World countries already resented buying seeds that often contained genes that were originally collected from their own wild gene resources. Secondly, to environmentalists this smacked of technological imperialism, increasing the dependence of these countries on advanced western technologies and contributing to their already massive debt crises. Thirdly, plants in the field could be taken and propagated at will. Once a crop was grown in the field, preventing the free dissemination of its seeds to other interested parties was likely to be well-nigh impossible.

One solution to this impasse might have been for the commercialization of the technology to be restricted to advanced western countries, which have the legal systems to support its exploitation through the contentious process of patenting of living organisms. Elsewhere, the new methods and products could be made freely available to Developing Countries as an act of largess. But even this presented difficulties.

Despite many bland assurances from expert committees, there were still fears about the environmental consequences of releasing genetically engineered organisms of any description into the environment. Their ecological impact was largely a matter of speculation and few experiments had been done which provided hard data about the possible environmental consequences. In Britain, Europe and America the new technology is subject to tight controls, a fact which was often cited to reassure a sceptical public, to illustrate the responsible way in which the technology was progressing from laboratory to supermarket shelf. In Developing Countries such rigorous environmental considerations were much less likely to apply. The proliferation of the nuclear capability and sophisticated weapons technologies illustrated how difficult it was to police the spread of new scientific developments, even with United Nations backing. How much more difficult would it be to police the use of genetic engineering, which had now become a relatively simple and routine laboratory technique?

So, in less than a decade plant genetic engineering manoeuvred itself into a commercial, political and environmental corner. It soon became apparent that marketing the technology might not be as easy or as quick as many people had been led to believe. Today, its public image is still deplorably bad and some multinationals have already resold their interests in plant breeding. The commercial returns no longer seem so assured.

Currently, plant genetic engineering is going 'green'. It has changed its name and slipped beneath the blanket of 'plant biotechnology', a less emotive, catch-all term which includes everything from the use of reeds in sewage processing to propagating garden plants by tissue culture.

The targets have shifted too, towards the replacement of environmentally damaging agrochemicals with genetically engineered, disease-resistant crops. Pest-resistant crops, carrying genes for naturally occurring insecticidal products, originating from sources ranging from tropical beans to bacteria, are now becoming available. There is no doubt that such plants will play a major part in reducing dependence on chemical control of pests. The new objectives that plant biotechnologists are setting themselves may do much to reassure the public.

Curiously, organic agriculture and plant genetic engineering may turn out to be highly complementary methods of producing quality food with less reliance on agrochemicals. But in the immediate future there are at least two major reasons why European agriculture will depend heavily on these controversial chemical products.

One is that a large-scale switch to organic methods would depend on a major change in public attitudes towards food. Almost twenty years ago I began my scientific career in a vegetable research station, engaged in a search for the perfect Brussels sprout. The designer sprout needs to be high-yielding and of uniform size and maturity, so that it can be harvested and graded mechanically and will freeze uniformly. At the same time as we worked towards the ideal sprout, the research station disbanded its food-tasting panel, which until that time had adjudicated on the question of flavour.

Although I didn't realize it at the time, the two events

probably signalled a turning point in public attitudes towards food. Low cost, convenience and appearance of the end-product now take precedence over all other criteria for the bulk of vegetables that are marketed to consumers. We have come to expect blemish-free food. If we were to have vegetable crops that relied on organic methods of pest control, we would have to accept a noticeable deterioration in their appearance, since such techniques are rarely as effective at controlling pests as blanket pesticide applications.

Consumer psychology would have to change. Instead of taking vegetables back to a supermarket and complaining if they contained a caterpillar, we would have to welcome the presence of the odd pest or two as a sign that the food was uncontaminated. The caterpillar would be an insurance, leaving us secure in the knowledge that if the pest could eat the vegetable and survive, so could we. Of course, some people can already accommodate this change in food psychology, but they represent only a small proportion of all consumers. It takes a major feat of imagination to envisage a time when every family will willingly accept a frozen caterpillar in its frozen sprouts.

But even supposing that a sufficiently large public demand existed, there is a greater problem. The farmers, who bear the brunt of public outrage over the excesses of their industry, are only the most visible part of a massive iceberg that is composed of the agriculture and food industries. Keeping them afloat, largely out of sight, are enormous transnational companies that deal in seeds, fertilizer and agricultural and processing machinery, directly or indirectly employing hundreds of thousands of people. The agricultural and food industries in developed countries have enormous built-in inertia to change. Any switches in policy have to be based on sound commercial principles. Organic agricultural methods quite simply mean lowering agricultural inputs, selling less of most of the commodities that the farmer currently uses. It is bad for business.

As a result, a major switch to organic methods cannot

happen quickly, but genetic engineering may offer a half-way house. It can be used to make crops resistant to pests and diseases in a way that could drastically reduce the amounts of agrochemicals that are used in the production process, but unlike organic agriculture it is a high-technology process which should be a much more profitable commercial proposition than organic technology. The benefits to a farmer of being able to plant a crop, safe in the knowledge that it is genetically immune to major pests and diseases and will not need to be sprayed, should be sufficient to ensure that he will pay a higher premium to grow it, provided that we will pay a realistic price to eat it.

Genetic engineering is a product of modern reductionist science, created by scientists who explore the molecular basis of life. The rationale is that by dismantling an organism into its component molecules and by understanding how these interact, it is possible to understand how the organism works. With that knowledge, the organism can be reassembled in a novel way, to our own advantage.

This Lego-kit philosophy of life has already produced tomato plants that contain a gene from a bacterium called *Bacillus thuringiensis*, or BT as it is known in the trade.

Organic farmers already use BT. They mix it with water, spray it on their cabbages and it kills caterpillars. Genetic engineers have taken this one step further, have isolated the gene for the toxin from BT that kills the caterpillars and have put it in the plant. Now the plant kills the caterpillar when it takes its first mouthful.

The gene product that does for the caterpillars is a protein that is harmless to humans, as far as is known, so all this seems safe enough. Genetic engineers and plant-breeding companies are confident that they can refine plant products like this even further. In seed crops they can even ensure that such genes are not expressed in parts of plants that are eaten by humans, if this should prove to be necessary. They believe that this approach will offer an effective and long-lasting crop defence mechanism against insect and fungal pests, although many crop pathologists who have spent a professional lifetime watching pests adapt to every new crop protection measure that human ingenuity can devise are rather more sceptical.

It would be naive to believe that plant genetic engineering will prove itself to be a completely benign science. History shows that all technologies have some adverse effects on the environment and in the case of genetic engineering there are many unanswered ecological questions. Closely related organisms can readily exchange genetic material and there is every likelihood that genetically engineered genes will move from domesticated species into their wild relatives. There they would have the capacity to spread and disrupt the equilibrium of ecosystems and food chains. It does not take much imagination to predict what would happen if insect resistance genes escaped from crop plants into wild species that are food plants for uncommon butterflies or other harmless insects, for example.

By manipulating DNA, the molecule which forms the chemical basis of our existence, and by reconstructing it in ways which are beyond the scope of natural evolutionary processes, we have taken a giant step towards ultimate control of our own destiny. This awesome power has already been learned and used with very little discussion of the ethical, moral and ecological consequences. Commercial forces have demanded that the technology must be applied with minimum delay. In the early years of the next century genetic engineering will be such an essential tool for matching crops to rapidly changing climate that there will be even less time and inclination to debate the long-term consequences.

We have to hope that this new technology does work and that it does prove safe and acceptable to the public. Its value in the long term, in helping plant breeders to produce crops for a rapidly changing climate, is incalculable.

The benefits it might confer on wildlife, by allowing farmers to grow crops with lower agrochemical inputs, are enormous. The very fact that plant genetic engineering could increase crop yields should also be of comfort to all those who love the countryside, since raising yields per hectare at a time when total production has become limited by financial and legislative constraints means that more food will be grown on less land. The challenge to conservationists will be to make sure that as much of the redundant land as possible is used for the benefit of the flora and fauna, and not for other forms of development.

SEVENTEEN

Wasted Acres

> If we can compare the impact of traditional farming on our flora to that of a frozen glacier in a state of retreat, modern agriculture behaves like a glacier in permanent, animated advance, and increasingly our contemporary flora is dominated by the limited number of adaptable species that can survive such conditions of turmoil.
>
> RICHARD MABEY, *The Flowering of Britain*

The farming industry is in deep recession. Before the end of the decade, six out of every ten British farmers will probably have gone out of business.

The industry is a victim of its own success and of economic, political and environmental circumstances. The need to reduce agricultural surpluses within the European Community and to establish new ground rules for world trade make it inevitable that the EC's complicated price support systems will disappear, together with many of the farmers who depended on them. Meanwhile, the coming period of climatic change is almost certain to compound the farming industry's predicament.

Farming is a business fraught with uncertainties, but few of the variables that farmers grapple with are as unpredictable as the weather. Financial success or failure can quite literally depend on which way the wind blows.

1990 was a traumatic year for many of Britain's cereal farmers. Their problems started even before the mild winter of 1989/90, which for the second year running promised to pave the way for major pest and disease problems as crops matured.

The long, hot summer of 1989 produced an early harvest in many parts of Britain. Modern farming is a hectic business and land which is not carrying a crop does not generate an income, so farmers were encouraged by the favourable weather conditions to sow their winter cereals early. For some it proved disastrous.

The combination of early sowing, an unusually frost-free winter, early maturing varieties and a mild spring, which allowed plants to develop faster than usual, produced winter barley and winter wheat crops that had begun to develop vestigial ears by the time savage late frosts struck in early April 1990.

Three frosts of exceptional severity were recorded on successive nights from 4th April 1990, a sequence that had only been recorded at the Heathrow meteorological station once before in the last forty years, in 1968. In many fields 90 percent of ears were killed by the extreme cold. For some crops there was hope of later flowering on regenerated side shoots, but for the worst damaged samples the only option was to plough the crops in and resow, in conditions that were made exceptionally difficult by an unusually dry spring.

In terms of accumulated temperature over the winter months and its affect on cereal development, the winter of 1989/90 led to the earliest spring since 1925 in some parts of the country, and the second earliest on record. The fate of the frosted wheat and barley illustrates the dilemma that will face farmers in a period when climates will become milder but less predictable. Calculating optimum sowing times will be even more of a gamble than it is at present.

The only options available to insure against a repeat of the 1990 frost damage are to sow later or to use later maturing varieties. But there are other factors which will influence both of these decisions. 1990 also turned out to be a second successive year of extreme summer drought. By the time farmers were ploughing in their frost-damaged crops, water restrictions had already been in force in some parts of the south-east for several weeks. As the season progressed, later sown crops which developed more slowly were the ones that suffered most severely from drought.

Even when harvests were successfully completed, the legacy of the drought persisted. Preparing satisfactory seed beds proved to be unusually difficult. After ploughing, hard-baked soils refused to break down into a fine tilth, despite more intensive (and more expensive) cultivation.

On the evidence of recent summers, it might seem that the summers of the twenty-first century will bring nothing but trauma for farmers. However, some agricultural scientists see things differently, arguing that the projected climate will actually improve crop yields.

Their grounds for optimism stem from the fact that carbon dioxide is the raw material that plants use to make cells, so plants will grow better if more of the gas is available. Photosynthesis, the elegant biochemical trick whereby plants convert gaseous carbon dioxide to sugars by using energy from sunlight, works more efficiently at higher temperatures. There is also evidence that plants use water more economically as carbon dioxide levels rise. The benefits of higher carbon dioxide and temperature levels can easily be demonstrated in the laboratory. The atmosphere in some commercial glasshouses is already routinely enriched with carbon dioxide to improve yields of tomato and other horticultural crops.

These convenient facts have been eagerly grasped by those who argue against moves to impose tight restrictions on carbon dioxide emissions and who oppose the introduction of drastic measures like carbon taxes. Some take the optimistic view that induced climate change might actually bring an unexpected windfall, inasmuch as it will boost crop yields in this way.

Closer scrutiny of the evidence for the beneficial effects of carbon dioxide enrichment is less comforting. Some studies show that the effect is transient and that although more of this essential gas does boost plant growth in the short term, the effect declines as the plant develops and other factors become limiting. Any idea that raising carbon dioxide levels might give us something for nothing, by increasing yields at no cost, is almost certainly nonsense because higher levels of other essential nutrients, like nitrogenous fertilizers, would also be needed. Significantly, most studies on the subject have been conducted in the confines of carefully controlled laboratory environments, where plants have not been exposed to other complicated stresses that also come into play under field conditions. Temperature and carbon dioxide levels which favour crop growth will also encourage weeds. Pests and diseases will benefit too and it seems certain that insect pests will increase. So the concept that atmospheric carbon dioxide enrichment might deliver an agricultural bonanza is questionable, to say the least.

The situation with respect to one of the great enemies of farmers, fungal pathogens, is harder to predict, although it is known that climatic conditions play a critical role in the spread of fungal diseases. There is no doubt that mild, wet winters will favour their growth and enhance the spread of fungal diseases that attack crops, especially if the same conditions promote the growth of host plants which carry them through from year to year, acting as a 'green bridge'. Hot summers can also favour some fungi, and powdery mildews were particularly prevalent in the dry summers of 1989 and 1990.

As any enthusiastic mycologist will quickly point out, fungi are extremely important organisms. The average British citizen's acquaintance with the group is limited to an occasional serving of the cultivated mushroom, and most of us remain blissfully unaware of the key role they play in our everyday lives.

Some species, like *Phytophthora infestans*, causative agent of potato blight, have changed the course of world history. When potato blight struck Ireland in 1845/6 it caused a famine which killed one million people and forced a further million to emigrate to the United States, leaving a political legacy that is still felt today. It caused the Conservative cabinet to repeal the Corn Laws, ushering in a free trade policy which precipitated the resignation of Sir Robert Peel, the Prime Minister, and triggered an agricultural decline.

'Rotten potatoes have done it all,' remarked the embittered Duke of Wellington. 'They put Peel in his damn fright.'

Such is the power of a fungus with spores that are only a few thousandths of a millimetre in diameter.

Potato blight infestations are predictable and are governed by particular combinations of temperature and atmospheric humidity. Warm, humid periods in summer favour its spread.

Several crop fungal diseases have very precise climate requirements. One such is ergot, a notoriously poisonous species which can often be found on forage grasses. It may become much less frequent in areas of the country that experience long dry periods in the early summer. Research carried out in the late 1950s showed that its incidence here is strongly correlated with high relative humidity and low maximum temperature in June.

Mediaeval peasants were familiar with the effects of this deadly species. The symptoms are horrific. Convulsive ergotism is manifested in uncontrollable twitchings and spasms and is accompanied by a numbing of the hands and feet, resulting in a sensation that has been described as being akin to swarms of ants running around just below the victim's skin. In gangrenous ergotism limbs become swollen and subject to violent burning pains, then gradually become shrunken and mummified, with the gangrenous extremities finally dropping off. Although the disease is a slow, progressive one, the first symptoms can appear within a day of eating ergot-infected grain.

In Britain there is only one recorded instance of gangrenous ergotism caused by eating infected grain, which claimed the legs of the wife and five children of a labourer in 1762. However, in 1927 an epidemic of mild ergotism struck two hundred Jewish immigrants in Manchester who had eaten bread containing a high percentage of ergot-infected flour.

The favoured host of ergot is rye, but in Britain it is most commonly found on forage crops, like rye grass. It has the same effects on grazing animals as on humans, so farmers would no doubt welcome a warmer, drier climate which would discourage its spread. A period of warm, wet weather at the time of infection aids the production and dispersal of spores. Grass florets, which are attacked by the fungus, only remain open for a short period, but in summers where flowering coincides with mild, humid conditions, the level of ergot production is high. The wet summer of 1988 was an 'ergot year' in my locality, whereas in the hot, dry summers of 1989 and 1990 the fungus was much harder to find. Continued dry summers will mean that its incidence will probably decline.

The factors that influence the final yield of crop plants are so many and are so diverse that it is impossible to predict with any confidence what the net effect of more carbon dioxide and higher temperatures will be. Scientists who predict that climate change will boost crop yields, basing their statements on evidence gained in laboratories under controlled conditions, should talk to the farmers who are familiar with the bewildering array of stresses that beset a crop between the time that it is sown and the time that it is harvested. Even hot weather can be bad for them.

In southern Britain at least, it seems likely that the trend in climate change will be to exaggerate temperature extremes at the upper end of the range, leading to higher summer temperatures. There is good evidence that some crops actually yield less as day-time temperatures rise above a comfortable level. A recent study on field pea crops in the United States showed that their yield begins to decline once temperatures climb above 26°C.

The evidence so far indicates that there are many complicating factors which make it impossible to predict any net beneficial effect of atmospheric carbon dioxide enrichment on the current range of crops that are grown in Britain. Abnormal and extreme seasonal fluctuations in weather conditions are likely to have a much greater impact on day-to-day farming operations than any gradual changes in crop performance.

*

Farmers are notorious for complaining about the weather, and in recent years the most common gripe has been lack of rain. That may well be an abiding mid-summer problem by the middle of the next century, but in winter too much precipitation could also be a major source of discontent, especially in the north and west.

If a pattern of dry summers and wetter winters becomes established, there will certainly be some unfortunate consequences for agricultural soils. After heavy rain, rivers take on the colour and consistency of oxtail soup, as they carry suspended soil particles towards the sea. The scale of soil erosion in Britain is trivial compared with that in many tropical countries, but it does, nevertheless, worry some geomorphologists and agronomists.

Any cultivation of soil tends to result in a gradual creep downhill, especially on steep slopes. Hedges and walls planted or constructed at right angles to the direction of slope contain this creep, so that over a period of time a distinct step can usually be seen between sides of a hedge, with an accumulation of soil on the upslope side.

Once hedges are removed, the downward movement of soil accelerates as rainwater run-off sweeps downhill, carrying cultivated soil with it. As the speed of the surface water increases, any cracks in the soil that are created by shrinkage during droughts are widened into gulleys. Compaction of soil by tractors wheels also channels water into rivulets and even small streams. Soon a steady drift of fine soil particles becomes a flood, and serious soil erosion starts.

The combination of intensive agriculture, creating prairie-like fields with no hedges to interrupt the flow of run-off, has made some areas of cultivation especially prone to erosion. This situation is bound to be exacerbated by any change in climate that would produce drier, baked soils in late summer, where surface run-off would be accelerated, and heavy rain in winter, which would increase the speed of erosion. Some of the most fertile, intensively cultivated soils could quickly be washed away.

An inkling of the potential scale of this problem comes from events which took place in Rottingdean, on the South Downs in Sussex, in the winter of 1987. A morass of muddy water was swept off ploughed fields by heavy rain, engulfing eighty houses and causing over one million pounds' worth of damage.

The damage to property which results from a local catastrophe of this kind represents only one facet of the overall problem, because fine silt is also washed into the drainage system, clogging drains and flowing into rivers, where it can smother the stream bed flora and fauna and destroy fish stocks. Silt is the largest single source of water pollution. The Mississippi was once described as being too thin to cultivate and too thick to navigate, a gross exaggeration which nonetheless contains a grain of truth.

Some of the worst instances of soil erosion and its related effects occur on newly restored opencast coal mining sites, where a combination of inherently poor soil drainage, no hedges and recently cultivated soil creates ideal conditions for rapid run-off and soil loss. After a heavy rainstorm it is not uncommon to find a river of muddy water running downhill from land which has been treated in this way.

So far soil erosion is a local problem in Britain, but a change in climate could make it a regional or national issue. Some of our most fertile agricultural soils could find their way into the river systems and ultimately into the sea.

Soil is a renewable resource which most biologists believe is grossly abused. It takes somewhere between two hundred and a thousand years for an inch of topsoil to form, depending on climate and soil type. This much soil can disappear in the space of a few months or even weeks from land which is left exposed to the full erosive powers of the wind and rain.

The large landowners who have followed the prairie farming route will find themselves with the choice of expensive preventative measures to control soil erosion or being faced with the spectacle of the best land being washed away, with the silt laden run-off creating an environmental mess which they may well have to pay for.

There are several remedies that may help to control the problem. Better land drains and ditching and also a trend towards replacing hedges on vulnerable land can be beneficial. Skilled cultivation, with sloughing across contours, also slows water run-off. So-called conservation-tillage, which breaks up the subsoil without turning over the topsoil, is even more effective. No-tillage methods, where herbicides clear land of weeds and new crops are planted into unploughed soil, can help in extreme cases.

These methods are becoming widely used in the United States, where farmers have first-hand experience of the worst effects of abuse of the soil and its consequences in climates where droughts are frequent. In the early years of this century a long sequence of dry years turned vast areas of the Great Plains into a dust bowl. In 1934 hot, dry winds whipped soils into the air and created dust storms that turned day into night. In May 1934 the entire eastern half of the United States was blanketed by a vast cloud of dust. The sky was filled with topsoil which had once underpinned the livelihood of Kansas farmers.

The dust bowl region is still blowing away, and future climate change is likely to make the situation even worse. Other areas of intensive agriculture are also badly affected by drought problems which make their soils vulnerable. California is currently enduring its fifth year of severe water shortage.

These grim experiences with soil abuse have left American agricultural scientists with a keen appreciation of the value of organic methods of soil cultivation. They assess their worth in terms of hard-nosed economics and do not treat proponents of organic methods with the same degree of condescending disdain that is so often the case in Britain. An extensive cost-benefit analysis of organic agriculture, emanating from Cornell University, has shown conclusively that if US agriculture cut inputs of agrochemicals by half and employed organic methods like crop rotation, which play an important part in soil conservation, then productivity would actually increase. Food prices would rise by less than one percent.

Organic methods of cultivation are an insurance against soil erosion. Organic farmers ensure that the land is covered with well-developed vegetation of some description during the wettest seasons of the year. Autumn sowing of crops should help in this respect, and if the land is not immediately used, temporary seeding with ley crops will be a wise precaution for farmers in high rainfall areas with land on sloping ground. Green manures – temporary crops which are cultivated over the winter period and then ploughed in – also stabilize the soil surface, locking up nutrients which would otherwise be leached out by rainwater. They add humus to the soil, which increases its water-holding capacity.

Without these soil conservation methods, the only way to shore up soil fertility is to use high inputs of artificial nitrogenous fertilizers. But heavy leaching by rainwater can make such measures an expensive waste of time and money. Nitrate in ground water has become an important environmental issue in Britain, and low-nitrate zones are currently being established in some areas of intensive arable agriculture. There is a trend towards severe restrictions on applying nitrogenous fertilizers in autumn, where risk of leaching is highest, and this will become even more important if wetter winters become more frequent. Fertilizer application is just one of many farming operations that will need to be retimed in the light of changing climatic trends.

Some ecologists are predicting that weeds will do rather well out of climate change.

Since one man's weed is another's wild flower, there has always been some difficulty in their precise definition. The word is usually taken to mean any plant growing where it is not wanted by man, a definition that can include anything from the daisy in the lawn to moss blocking a gutter.

Weeds have had a chequered history and have blighted

man's efforts to cultivate bare soil since the beginning of agriculture. Many of the most familiar species are annuals which naturally colonize disturbed soils, a habitat which would have been relatively scarce five thousand years ago, when much more of the landscape was covered with forest. The primitive agricultural techniques of Neolithic farmers, pushing back the frontiers of woodland and creating bare soil, were as beneficial for these vegetable opportunists as they were for the crops they cultivated. Annual weeds and agriculture are inseparable.

Successful weeds are supremely adaptable plants, capable of germinating and growing more quickly and vigorously than the crops they compete with. For them, changes in seasonal weather patterns will surely make Britain a land of new opportunities. During periods of severe drought many of the more delicate, shallow-rooted plants will die, creating areas of bare ground. Poor emergence and establishment of droughted crops will similarly leave large areas of ideal habitat for weed colonization.

Some research by Ceri Evans and John Etherington at the School of Pure and Applied Biology at the University of Wales gives some clues as to how weed species might perform in a drier climate. They investigated the germination potential in dry soils of a range of British plants, demonstrating that curled dock (*Rumex crispus*), already the bane of many a farmer's life, has a remarkable ability to germinate in soils that are dry enough to cause most crops to wilt permanently. This species is also known to have a very long period of seed viability, with 52 percent germination being obtained from seeds which have been buried for half a century. These seem to be ideal credentials for thriving in the new climate.

Two other species turned out to be almost as tolerant of moisture stress. These were the green-flowered wood sage (*Teucrium scorodonia*) and dyers' rocket (*Reseda luteola*), whose tall yellow spikes of flowers are a common sight on waste ground, railway sidings, river banks and rubbish tips. The significant feature of these two species is that they are not presently considered to be serious agricultural weeds, but their ability to germinate in dry soils makes them contenders for the title in a drier future.

This ability of seeds to germinate in conditions that would kill surrounding vegetation has some interesting ecological implications. It may be that prolonged water stress after germination would kill the seedlings before their roots could penetrate to deeper, moister soil but, provided that they can survive, it is equally likely that such germination behaviour would allow them to establish a bridgehead as primary colonizers in bare soil. Given that curled dock, wood sage and dyer's rocket all have a high seed output, the chance of occasional successful establishment would more than compensate for regular failures. Such unstoppable germinaters might well be the weed shock troops of the new agricultural environment.

One other fascinating aspect of Evans' and Etherington's germination experiment concerns the response of marjoram (*Origanum vulgare*), an aromatic herb that is a frequent plant of dry sites, but by no means a weed. In soils which were drying out, these seeds imbibed water, swelled and burst their seeds coats, even exposing the embryonic root. Then they remained dormant until water became available again, when they quickly burst into life. This two-step germination behaviour, where the seeds have been primed to take instant advantage of available water, was observed twenty years ago in seed germination tests carried out in Israel, where drought is a perennial, national problem. A similar phenomenon has also been described for some dry-season grass species in New South Wales in Australia.

No one knows how widespread this kind of germination response is in British wild plants, but developing an understanding of it is important for two reasons. One is directly related to the effect it might have on weed establishment, the other is that it might be useful to prime seeds of crop plants in this way, to aid their germination. Priming seeds of horticultural plants is already carried out on a commercial scale.

We clearly need to know more about weed seed germination in dry soils and it seems certain that we will also need effective and environment-friendly means to control the spread of weeds. One promising approach, being pioneered at Lancaster University, involves using fungal herbicides. Strains of fungi, which are harmless to crop plants but lethal to weed hosts, are being developed. This is part of a growing trend towards the use of biological methods of pest and disease control, supplementing and perhaps eventually replacing some of the more dangerous chemical methods that are now in use.

In the British landscape, the heyday of the weed probably occurred between the beginning of the nineteenth century and the beginning of the Second World War, when mechanized agriculture allowed cultivation on a grand scale and before effective chemical herbicides came into widespread use. The age of steam had created the means to plough large areas of land more quickly, but manual weeding was still the order of the day. During this transitional period the landscape became a source of inspiration for artists, and cereal fields with drifts of poppies, cornflowers and corncockles became a popular artistic cliché.

In 1925 A.H. Church wrote a delightful, descriptive and botanically impeccable account of the plant life of the Oxford district through the seasons, which gives a vivid impression of arable crops in the days before intensive use of agrochemicals changed agriculture and the landscape so drastically. He described overgrown bean fields, which were notorious for their weed flora, as being ablaze with poppies, corn gromwell, corncockle, corn marigold, bugloss, wild radish, mayweeds and melilot. 'Children and adults, ranging the countryside, collect bunches of ox-eye daisies, ragged robin, sainfoin, quaking grass, yellow flag. . . ' he commented, in passing. The scene he described, a naturalist's idyll, was swept away by the advent of chemical herbicides.

Today, the commonplace agricultural weed flora of Church's Oxford has all but disappeared. The plants he described with such obvious affection epitomize a lost age of agricultural innocence, which corncockle and cornflower have come to symbolize. In some ways they are strangely inappropriate motifs.

Both owed their presence in the landscape to agriculture. Corncockle is an alien which was often introduced as a contaminant of cereal seed imported from Europe. Cornflower is a native plant, but the drifts of 'bluebottles', as they were known in Victorian times, were again the result of seed contamination of imported European cereals.

The nostalgic fondness for these long-lost arable weeds, which owed their existence to the agricultural industry, has arisen not because they symbolize lost nature, but because they represent a particular kind of artificial, agricultural landscape.

Our preconceptions of what rural landscape should look like are shaped by the work of artists and writers who drew and wrote prolifically about rural Britain at a particular point in history. This was a narrow window in time, framed by a preceding period when the countryside was a wild, inaccessible and threatening place, and an era of prosperity, when the romantic landscape and simple rustic charms were highly valued by a wealthy and increasingly mobile middle class, who could observe the countryside from a comfortable distance at their leisure. This idyllic landscape was managed with manual labour provided by a large rural population, so changes still unfolded at a pace which allowed wildlife to adapt. The time when the use of efficient machinery could bring about sudden and dramatic convulsion in the countryside was still some way off. In general, agricultural activities did not present an obvious threat to the native flora and fauna of Britain.

All that changed after the Second World War. The rapid mechanization of agriculture brought about a revolution in the speed and scale of everyday agricultural operations, while the use of intensive chemical methods compressed the work of days and weeks into hours. Suddenly, the countryside began to change so rapidly that wildlife could no longer adapt. The flora and fauna went into retreat.

At first, public attitudes to the changes were favourable. In the twentieth century, wartime exigencies, when every useful acre had to be cultivated, made people acutely aware of the need for adequate and reliable food supplies.

In an editorial in *Wild Flower Magazine* in September 1941, Edith Dent, President of the Wild Flower Society, wrote:

> In England the war has caused a vast improvement in the cultivation of the land – every corner, in time, will be used. It is hoped that the farmers will continue their good work in peace-time and will not sink again into idle and unthrifty ways.

They certainly did continue the good work, to devastating effect. Half a century later, it is clear that agriculture has done more than any other human endeavour to destroy the wild plants that the Wild Flower Society prized so much. Corncockles and cornflowers have gone.

EIGHTEEN

The Lie of the Land

> We should realize that the present interglacial warm period could be regarded as a fever for Gaia and that left to herself she would be relaxing into her normal, comfortable for her, ice age. She may be unable to relax because we have been busy removing her skin and using it as farm land, especially the trees and the forests of the humid tropics, which are otherwise among the means for her recovery. But also we are adding a vast blanket of greenhouse gases to an already feverish patient. In these circumstances Gaia is much more likely to shudder and move over to a new stable state fit for a different and more amenable biota. It could be much hotter or much colder, but whatever it is, no longer the comfortable world we know.
>
> JAMES E. LOVELOCK, *Gaia*. The first Leslie Cooper memorial lecture to the Marine Biological Association of the United Kingdom, Plymouth, 10th April 1989

We are on the threshold of change. As Britain becomes warmer, some plants and animals will inevitably disappear and new ones will arrive. Some much-loved features of the countryside may well be transformed almost beyond recognition in the space of a single human lifetime. The extra one or two degrees centigrade which the current round of climate amelioration will bring with it will be sufficient to bring the simmering problems which beset our flora and fauna to the boil.

It is difficult to know how to respond to this disconcerting prospect. My instinctive reaction, as someone who spends much of his life studying the current flora and fauna, is to be appalled that society's profligate exploitation of the environment has precipitated a climate change which will deface and possibly even destroy a natural resource which gives me so much pleasure.

But I have also had the benefit of a scientific training. This tells me that my subjective assessment of the situation is quite illogical. It tempers my instincts and warns me that if I view the transformation dispassionately, with cool, scientific logic, I might well come to the conclusion that there is little to get upset about.

The landscape has been changing continuously under the influence of natural shifts in weather patterns ever since the retreat of the glaciers at the end of the last Ice Age: the new climate and the flora and fauna that can adapt to it will be different, but not necessarily worse, than the current one. More recently, human interference has controlled the form and content of the landscape, quickening the pace of change throughout recorded history. The inhabitants of our islands have progressively left their mark on the landscape, to the point where there is now virtually no tract of land left which could be said to be natural, in the true sense of the term. Today, 'natural' landscape is only as natural as we allow it to be. So what does a little more change, which is well within the temperature range of past climate fluctuations, really matter?

The climatic shifts which will occur during the next half century will accelerate and extend a pattern of change which has, until relatively recently, been accepted as an inevitable consequence of progress. Provided that the climate stabilizes at the end of the current period of warming, we might even consider the next half century of changes to be generally benign or even beneficial. Agricultural technology should be able to cope with the new conditions, introducing new crops and new techniques to modify those which we currently depend on.

Most of us would welcome a climate that was warmer in

summer. We might be forced to make some adjustments to our lifestyles and we might need to invest large sums of money in reorganizing an increasingly unequal distribution of water, but many of these contingencies might be balanced by economic windfalls, like a reduction in energy use in a warmer climate.

It is even conceivable that wildlife might benefit. The diversity of flora and fauna in Britain is naturally limited, partly by the fact that the British Isles were separated from the Continental land mass before many species could cross the temporary land bridge, and partly because our seasonal climate is inhospitable to species from further south that need higher temperatures. Many of the rare species of plants and animals in Britain are rare precisely because they are at the northern edge of their range. In Europe, exotic rarities like hawk moths and hoopoes are much commoner. I have to admit to a flutter of excitement at the thought of bee-eater colonies becoming established on the south coast or at the prospect of hearing cicadas in the trees in the Midlands.

It seems quite a reasonable argument that a change in climate, which opened up new habitats in the south and pushed existing species northwards, could actually result in a national increase in species diversity, as southern species arrived on our shores. Gains could easily outweigh losses.

Some might argue that this artificial enrichment of our wildlife would be unnatural, but this is really a question of perception. The whole concept of naturalness, both in terms of landscape and in terms of distribution of the fauna and flora in Britain, is profoundly dishonest. There is little that is natural about the persistence of our native flora and fauna. It is increasingly managed and depends for its continuity on the concessions we make to it. Even the future of plants and animals that lead a seemingly undisturbed existence on nature reserves depends on continuous human interference to maintain habitats in a condition that will support their exacting environmental requirements. Conservationists spend a great deal of time and energy clearing scrub and preventing the natural reafforestation of areas where trees need to be excluded. Several of our most highly-prized wildlife habitats, like hay meadows, downland and heathlands are the result of centuries of human interference, mainly as a result of agriculture.

So it is not particularly easy to say with any degree of scientific, objective judgement that climate amelioration and the changes that it will bring to our countryside are changes for the worse. Anyone with a scientific inclination might actually derive a certain excitement and satisfaction from the opportunity to witness the effects of such a gigantic experiment, which promises to provide a great deal of interesting information about how plant and animal communities respond to climate.

I should be reassured. But I'm not.

I have to take science and objectivity with a pinch of salt when the future of our countryside is discussed, since it is the ruthless appliance of scientific knowledge that has been primarily responsible for transforming it into what is often biologically dull landscape. Relying on strict scientific criteria for deciding future strategy for responding to changes in wildlife and the countryside seems to be akin to asking the criminal fraternity to design burglar alarms.

I have to say at this point that by the word 'science', I do not mean 'knowledge ascertained by observation and experiment, critically tested, systematized and brought under general principles', as my dictionary defines it. Today, science is much more than this.

The popular media image of a scientist, as an independent member of that elite band of individuals whose job it is to study and interpret the world, is as outdated as equating the ideals of amateur sportsmanship with the pursuit of professional sport. Today there are almost no independent scientists. They are all contracted to employers, who direct their research either by controlling the flow of funds or rewarding performance.

Modern scientists are generators of wealth, and even in

the academic world the ethics of the marketplace are taking over. Unfettered exchange of information, once a cornerstone of academic life, is now restricted by secrecy agreements. Scientific method is as objective as ever, but the motives behind almost all scientific research go far beyond intellectual curiosity. It is an information-gathering exercise for economic development.

So I have qualms about allowing science to dictate policy on the imminent changes. Although I can claim to be a practitioner, I also harbour a deep distrust of scientific motives.

Instead, I prefer to view the coming problems with a broader perspective. The transformation of large areas of the Home Counties by the onset of a Mediterranean-type climate will certainly make many people – perhaps the majority – recoil in horror. The reason for this is that our landscape has a powerful emotional appeal which reaches far beyond scientific or economic objectivity. The landscape of rolling downland, fields, hedges, copses and woodlands is part of our natural heritage; part of our national identity. It is the backdrop against which much of our history has been enacted and figures prominently in our literature, music and art. If we are no longer able to experience the melancholy of a salt marsh, the setting for George Crabbe's poem 'Peter Grimes' would become an abstraction. If Dorset heathlands disappear, Thomas Hardy's novels would no longer have a living context.

If I am uneasy at the prospect of the hop fields of Kent being replaced by vineyards, tamarisks replacing hawthorn in hedges or the downland of Sussex slowly turning into dry scrub, this is not because I harbour unhealthy instincts of biological xenophobia. Such developments might well enrich the countryside in some respects. What I find alarming is that these small beginnings will be symptoms of something much larger. It will be the speed and irreversibility of climate change that will be so dangerous. The new round of disruption will accelerate changes wrought by intensive agriculture and urban development that are already viewed with growing alarm. It will finally wipe out any prospect of repairing much of the damage which has already been done to our flora and fauna. We will be past the point of no return.

So we must prepare ourselves for what will be, to begin with at least, another damage limitation exercise.

It has to be accepted that some of our wildlife and some features of our countryside will inevitably be consigned to the pages of history. It will be important to document these upheavals, since they represent a momentous step in the development of our culture. They will represent the first occasion on which we have succeeded in altering the environment for our whole flora and fauna at one fell swoop. Until now all the changes have been piecemeal and more or less localized, but this time the problem is infinitely worse than negligent burning of hedges or ploughing up of Sites of Special Scientific Interest. This time we have changed the ground rules which govern the very existence of whole classes of living organisms.

We must use knowledge gained by monitoring current changes to help formulate policy for preventing further, catastrophic consequences of uncontrolled and irreversible climate changes, which will occur if drastic steps are not taken to regulate carbon dioxide emissions into the atmosphere. The best outcome that can be hoped for from the current round of climate amelioration is that it will serve as an obvious and unambiguous warning of what will happen if we do not modify our attitude to the environment. The current phase of climate change is a serious but treatable illness with inevitable after-effects, but it could become terminal unless medication is applied.

There is likely to be considerable debate about who should monitor the changes as they take place. Requests for funding for this type of exercise from government coffers will probably often fall on deaf ears. The current trend is to target biological research on the basis of a cost-benefit analysis; this would be well-nigh impossible for an exercise that involved painstaking recording of unpredictable events, for an unspecified length of time, for no purpose other than the acquisition of potentially valuable knowledge.

Government-funded research into the effects of climate change will probably be targeted at problems that have practical, short- and medium-term economic implications. It seems likely that state-funded research will be preoccupied with technologies, like genetic engineering and developments in water management, which will help to combat detrimental effects of climate amelioration.

It is often difficult to define the immediate economic benefits of research into natural ecosystems, especially when it involves painstaking collection of data over long periods. The results of such research, where they successfully present data on long-term trends, can frequently be economically and politically subversive. Long-term studies on pesticide residues, for example, have been instrumental in securing the banning of many of the most noxious agrochemicals, like DDT, which were once classified as safe on the basis of the best testing procedures that were available at the time of their introduction. Long-term studies on the environmental effects of climate change would certainly generate equally expensive and embarrassing data. Governments will need to think very carefully about sanctioning certain kinds of environmental research, for fear of what might be revealed. Climate change will most certainly expose the woeful inadequacy of past measures to conserve our dwindling wildlife.

For these reasons we cannot depend on conventional science to record the fine details of the changing environment. It will be too busy dealing with the economically important consequences. So much of this work will fall to Britain's amateur naturalists.

There must be hundreds, perhaps thousands, of natural history societies throughout the country which are part of a tradition that dates back to at least the seventeenth century, when collecting natural history specimens blossomed into a respectable social activity. Some natural history societies grew to be large and influential organizations, and their researches appeared in the pages of learned journals. The onus for recording the detailed changes in the fauna and flora during the onset of the new climate will fall on non-governmental conservation bodies of all types, like the Royal Society for the Protection of Birds, the Royal Society for Nature Conservation, the Botanical Society of the British Isles and countless smaller but similarly motivated organizations. With a membership that includes many gifted amateur biologists, they can make a major contribution to recording the effects of climate changes. The challenge will be to find the funding and to provide the necessary organizational framework.

The revolution in information technology has provided naturalists with extraordinarily powerful methods for recording and storing data. The founding fathers of natural history, like Gilbert White, established the tradition of the diarist, which persists to this day in the pages of several national newspapers. The diary has remained the most common format for the amateur naturalist's records, describing day-to-day changes in the natural surroundings. Even Charles Darwin recorded many of his experiments in this form.

Such diaries have a literary quality as well as scientific substance, but they are notoriously difficult to extract information from. The advent of cheap, powerful personal computers with easily mastered databases now provides naturalists with the means systematically to record information and to recall it instantaneously as evidence of change.

Such recording of our changing flora and fauna will be an essential part of our response to climate amelioration, although it will do nothing to forestall the changes that are taking place. That will require practical measures.

The genie has been let out of the bottle; our industrialized society has precipitated a climatic shift which we can live with but cannot reverse within a timescale that has any meaning. The planet's thin green veneer of vegetation and all the life that it supports will change.

Armed with the limited knowledge at our disposal, should

we try to anticipate and prevent the worst effects, or should we accept them with equanimity and react to changes as they become apparent? When it comes to conserving our wildlife, we have two options open to us.

The first is to go in for wildlife gardening on a grand scale. We could do what we have always done and approach the problem with a siege mentality, performing piecemeal rescue attempts on species and communities that are threatened by changing weather patterns. If we were able to predict changes that will affect our wildlife with any degree of certainty, it might seem reasonable that we should take steps to conserve present species and habitats whenever possible. One obvious measure would be to move species that are under threat from changing local climate to other, more favourable, parts of the country. In most instances this would mean trying to recreate present-day south-eastern habitats in more northerly parts of Britain or moving species along altitude gradients. Even this apparently simple measure is fraught with difficulty.

There have been some notable examples of success in rescuing single species that are threatened by industrial or agricultural development. Heath fritillary butterflies were staring extinction in the face in the early 1980s, largely through the loss of their coppiced woodland habitat. Only thirty-one colonies remained in eight isolated localities. Traditional coppicing methods create ideal conditions for the spread of common cow-wheat, the heath fritillary caterpillar's food plant, and by intensively managing selected woodlands and carefully reintroducing isolated populations of the butterfly into new locations in Kent and Cornwall, its numbers have increased and its immediate future is now much more secure.

Otters too seem to be making a limited comeback, following determined efforts to conserve their habitat. Rumour has it that even Britain's rarest and most exotic plant, the lady's-slipper orchid, might be on the increase. Once reduced to a single, jealously guarded specimen in Yorkshire, it has been rescued by modern tissue culture techniques and is being reintroduced into the wild.

SPRING FEVER

Carefully nurturing a single rare species in a suitable habitat can be difficult but it is feasible. Currently English Nature, the successor to the Nature Conservancy Council, is developing a series of recovery programmes aimed at improving the status of our rarest plants and animals. This may be successful for individual species and will certainly generate a large body of valuable knowledge on their habitat requirements. But preserving whole ecosystems of plants and animals that are threatened by climate change is a problem which is infinitely more complex.

In 1964 work began on the construction of a dam in Upper Teesdale which flooded over a thousand acres of Pennine uplands and threatened a unique community of plants which had survived there since the last Ice Age. These formed the famous Teesdale flora of arctic-alpines and included the Teesdale violet, the azure-blue spring gentian, the dainty pink Teesdale primrose and the alpine meadow rue, together with a group of sedges and rushes that were less well known but were of equal or even greater botanical importance.

When the conservationists' appeals had been turned down and the bulldozers began to roll, a rescue bid was set in motion. As many as possible of these botanical gems, soon to be submerged, were uprooted and their seeds collected. At the same time, attempts were made to rescue and transplant two complete sets of plants to the Botanic Gardens of Durham University and to another site at Jodrell Bank in Cheshire, in an effort to maintain them in cultivation.

Soil types at the rescue sites replicated the conditions of Upper Teesdale and watering regimes which mimicked the natural conditions were also maintained, but nothing could be done about the change in climate. The experimental sites had a longer growing season and were consistently 2 to 3°C warmer than Upper Teesdale throughout the year.

Almost ten years later the performance of the transplants was assessed and it became clear that the change in climate had had a major effect on the competitive balance between the species. In the milder conditions, plants like wild thyme, harebell, rockrose, variegated horsetail and dog violet, which coexisted with the Teesdale rarities up in the high Pennines, became aggressive weeds, smothering the rarer and more delicate species. Constant removal of the lowland weeds of horticulture also became essential, otherwise they too would have swamped the transplants.

One by one, the transplanted rarities went into decline. The upland population of sea pink succumbed to an attack by moth larvae. Spring gentian did well initially but in the milder climate it appeared to produce flowers and seeds at the expense of vegetative growth and literally flowered itself to death. Teesdale primrose also began to fade away, as did several other rarities. Today, the Teesdale collection at Durham University Botanic Gardens is completely overgrown and no trace remains of any of the rarest plants.

The Teesdale transplant experience demonstrates the difficulty of any attempt to relocate whole plant communities, especially when they contain vulnerable species which are at the limit of their natural range. Even maintaining individual species in cultivation demands constant, intense management of weeds and unfamiliar pests. Together, the Teesdale assemblage of species no longer behaved as a balanced community when climate changed; competition soon eliminated the weaker-growing, and generally more desirable, species.

If we were to attempt to respond to climatic change by transplanting plant communities, we would need to perform the task on a far greater scale than was attempted in Teesdale, which was little more than a small-scale pilot scheme to maintain the rarities in cultivation. In a warmer Britain we would need to repeat the experiment hundreds or perhaps thousands of times.

If we erred on the side of prudence and began now, we would run into the problems of current climate changing the competitive balance between transplanted species. We would need to monitor and tend the transplants intensively in an attempt to ensure their survival until climate change caught up with them and hopefully restored some measure of balance.

If we wait until threatened plant communities are on the verge of extinction as a result of climate change, it will be too late to learn from mistakes and adjust transplanting techniques, should things go wrong.

In reality, moving plants and animals around in this way on a large scale is a desperate measure which would be impossibly expensive and irresponsibly ineffective. It may work for some specific cases, such as butterflies or grasshoppers, where they might be established in past habitats or even new habitats wherever suitable food plants are available, but in general large-scale direct intervention would not be an efficient use of limited resources. There is a better alternative.

Climate change has arrived at a time when pressures on the countryside are changing dramatically. The decline of the farming industry and current rapid advances in agricultural technology offer what may be the last opportunity to adopt a radical approach to coexisting with our wildlife. Perhaps it is time to abandon any pretence that we are actually living in a natural ecosystem; now we should begin to approach wildlife conservation with the same methodical management strategies that we apply to farming.

When climate changes, many of our nature reserves and Sites of Special Scientific Interest will be located in the wrong places. Those islands of habitat that we have done so much to nurture will have a bleak future. In order to respond to change, their occupants will need an escape route. They need corridors.

The provision of wildlife corridors is a well-established concept. You really need a kestrel's-eye view of a city to appreciate their value. Within any conurbation there are parks, green spaces created by cemeteries, derelict land, open grassland around industrial estates and leafy suburbs with tree-lined avenues. Each urban habitat is isolated, surrounded by concrete, glass and asphalt, and because of this most of the plants and animals that exist in these green islands are those that have been hardy enough to cling on as the concrete walls have grown up around them, those that have been deliberately introduced or those animals like pigeons, rats and starlings, that benefit from the effluent society in cities.

Similar but far richer pockets of wildlife exist in the open countryside, but instead of concrete, they are often surrounded by a minefield of intensive agriculture. The barriers are different but the fundamental problem is the same.

In cities the only routes open for natural colonization of the isolated green spaces are the green ribbons that run into cities' centres. These are the railway lines and arterial roads, which act as convenient corridors for animals like foxes and for song-birds. From the window of a railway carriage the embankments along an urban railway line are often a riot of colour. They are linear wildlife habitats, that sometimes widen into railway yards, often colonized by opportunist plants that are host to a wide range of insects and butterflies. The verges and embankments of roads and waterways perform the same function with varying degrees of diversity. These are the arterial routes for plants and animals, the only avenues that are open for movement in an inhospitable environment.

Some local authorities and conservation organizations make strenuous efforts to maintain and encourage this green wildlife circulatory system. Tyne and Wear, for example, is nurturing a well-thought-out system of strategic and local wildlife corridors and small wildlife links that maintain continuity between habitats, helping to ensure their survival. Corridors can range in size from the banks of major rivers to minor disused mineral railways, but their essential function is the same: to secure the future of populations of plants and animals by creating a spiders' web of habitat.

Such a system is desperately needed on a regional or even a national scale in the wider countryside, and the crisis in agriculture provides a golden opportunity to implement it.

With the decline of farming, large areas of land are about to become redundant. In some cases farmers are already being paid to keep their fields out of production, but there is still pressure for it to be put to good use. Wildlife conservation could offer a much better alternative than caravan sites and golf courses.

If a strategy was devised to use these areas of surplus agricultural land to link the present islands of intensively conserved wildlife, it would help to provide the continuity of habitat that is the mark of a wildlife corridor. The more mobile organisms could use these escape routes during the period of climate change.

Wildlife corridors created in this way would demand very intensive habitat reconstruction to restore the levels of species diversity that agriculture constantly works to eliminate, but I know that it can be done. Having been involved in the latter stages of the management of a site of old gravel workings, which have been transformed from a virtual moonscape to a well-wooded wetland habitat and are now designated as a Site of Special Scientific Interest, all in the space of a little over twenty years, I have little doubt that in many parts of the country the results of half a century of agricultural excesses could be reversed.

Current changes in the countryside already hold out some promise for the development of extensive corridors. A dozen community forests are planned throughout the country and if these come to fruition they would offer a link between existing fragments of woodland. Some of the best remaining parts of the landscape have also been designated Environmentally Sensitive Areas, and State payments are already made to farmers who use methods that ensure the survival of existing wildlife. The rudiments of a national network of wildlife corridors are already falling into place. With some imagination and the requisite political will a national network of wildlife corridors could be set up and their provision could become an inseparable part of the planning process for future land use.

The concept of 'green gain', where every type of economic exploitation would be balanced by the establishment of wildlife habitat, must be grasped by all planners. It may be too late for this to have much impact on the current pattern of change, but it would act to stabilize the countryside and its wildlife against future disruption.

A national network of corridors of this type, where broad ribbons of wildlife habitat would even run through the centre of housing and industrial developments, would have more than just biological value. They would be an essential step in reintegrating the population and its economic activities with the 'natural' environment. This would be an important move towards reversing our compartmentalized concept of the countryside: a concept where plants and animals have been squeezed into wildlife ghettos which are euphemistically termed nature reserves. This is the traditional strategy of wildlife apartheid which has made our flora and fauna so vulnerable to rapid climate change. An integrated approach to economic development and wildlife habitat would be a much healthier way to coexist with what remains of our 'natural' heritage.

For the vast majority of us, the worst aspects of environmental degradation do not impinge on our everyday life. We can still take a drive into the countryside and see rolling green fields and woodlands. They look fine, unless you examine them closely and talk to those who remember what they used to be like and know what has been lost. We can still walk along the wide, sandy beaches of the East Coast, and apart from the odd patch of oil and some unsavoury flotsam on the strand line, that looks pretty healthy too. Environmentalists have campaigned for years about the deteriorating state of the North Sea, but unless you are a trained marine biologist or happen to earn a living from fishing in the North Sea, this large area of grey and often stormy water looks reasonably healthy all the way to the horizon.

Nevertheless, almost everyone is vaguely aware that something pretty unpleasant has happened to a great deal of the wildlife that could once be found in our islands. But somehow, it has all become far removed from everyday experience.

The fact is that most of us never encounter many of the species of fauna and flora that have suffered the onslaught of five decades of industrial growth and intensive agriculture. And that is really the heart of the problem. Much of our wildlife is inaccessible, furtive, or so rare as to be out of reach of all but the most dedicated naturalists. Unless you happen to be particularly obsessed with smooth snakes or mole crickets or the oblong-leaved sundew, and are prepared to search them out in their few remaining haunts, it is unlikely that you will ever see them. They are the concern of the minority.

We have become an urban society, which for the most part has lost touch with the intimate detail of the natural world. Most of us experience natural history vicariously, through spectacularly televised wildlife programmes interpreted by expert presenters. If anything, televised natural history has distanced the population from the natural environment. Why spend six hours in a cold, wet hide with only a remote chance of glimpsing a badger when you can see the intimate details of its domestic life from the comfort of an armchair?

The ease with which we can watch and admire plants and animals through the medium of television masks the fact that it is becoming increasingly difficult to find these species in the wild. Television and a lifestyle which is remote from the natural world have shielded us from the havoc that has been wrought amongst our wildlife.

So an approach to using land whereby wildlife habitats were no longer part of the bureaucrats' compartmentalized concept of planning would help to reintegrate the flora and fauna into everyday life. If society is to care about what happens to the wild plants and animals, they must be accessible.

This book of full of ifs and buts and maybes. Many scientists will probably loath it. There is still a hard core of sceptics who refuse to countenance any response to the problem of climate change without continually demanding more experimental evidence. They want to see hard data on real, impending disaster and dismiss any speculative approach to the effects of climate change that is based on combining straightforward observations, past experience and projected change as 'bad science'.

The irony inherent in this attitude is that some technologies that have wrought wholesale damage to the environment have been introduced on the basis of a similar combination of criteria. But for reasons of economic expediency, there are limits to the extent of testing of such scientific advances.

For example, we routinely test the safety of our agrochemicals by the best methods that are currently available, over what is considered to be a prudent time interval, before they are released for general use. And yet, as knowledge of scientific methods improves it is not uncommon for pesticides that were once classified as safe, such as DDT, to be withdrawn from use when new health or environmental risks are discovered. There is a balance between economic benefit and caution, which many would say is weighted too heavily towards economic benefit.

Those concerned with driving economic progress forward are not gifted with any greater degree of foresight than those who are concerned with the looming environmental crisis. It seems unreasonable, to say the least, that those who question the direction in which economic progress is taking us should need to support their views with much more scientific evidence than those who are introducing the new technologies aimed at generating wealth, but such is often the case.

Despite the fact that climate amelioration was predicted twenty years ago, no one thought to carry out experiments to gauge the effect of higher temperatures or a carbon dioxide-rich atmosphere on the plant communities of a hay meadow or downland, or the marine life in shallow waters. Such investigations have not figured in national priorities when energy or transport policies have been formulated.

The amount and quality of scientific data which the establishment demands is tempered by economic or political expediencies. Technological progress that has obvious short-term benefits is often applied after the minimum acceptable scientific experimentation. When environmental concerns about technological progress threaten its implementation, much more rigorous scientific supporting data is always demanded. Technological progress operates under the same rules as criminal justice and is presumed innocent until proved guilty. In this context, science operates under the same rules as the adversarial system in a court of law. Our present predicament is the result of too much technology and too little science.

With respect to the wider environment, the fundamental basis of economic decision making is an extreme example of 'bad' science, based on a process which is the complete reverse of sound scientific procedure. Now we can only attempt to predict the probable effects of climate shifts on the basis of scanty evidence and wait, watch and measure the changes as they occur.

Fortunately, there are some signs that establishment attitudes to the problem of climate amelioration seem to be changing. They are advancing slowly, too slowly, beyond the position where irrefutable evidence of damage is demanded before any remedial action is taken. Now that the reality of climate change is accepted, the simple logic of the precautionary principle, where the environment is always given the benefit of the doubt, is gaining adherents.

It has been customary, throughout the history of environmental debate, for those who voice concerns over environmental issues to be dismissed as cranks or eco-doomsters. They have consistently been accused of shouting 'fire' before anyone could smell any smoke.

Common sense tells us that experimenting with such a complex system as the environment, without a proper understanding of its mechanism, is an extremely risky way to live. If someone shouts 'Fire!' in an office building, the sensible course of action is to close all the fire doors until you are sure that it is a hoax, rather than wait until you can see the flames advancing down the corridor.

Index

Abax, 29
Abeliophyllum distichum, 20
Achillea, 34
Adder's tongue fern, 136
Adventive Flora of Tweedside, The (I.M. Hayward), 59
Agricultural landscape, changes in, 180–90
Agriculture, *see* Farming industry
Agrobacterium tumefaciens, 195
Agrochemicals, 196–7, 199, 216, 225, 235
Agryranthemum, 16
Alfalfa, 44
Algal blooms, 159–60
Alien plants, introduction of, 58–62
Allium, 21
Alpine meadow rue, 228
Alpine milk vetch, 57
Alpines, 41–2, 43, 45, 55–7
Aluminium acetylacetonate, 30
Aluminium salts, 30
Amazonian waterlily, 155
Ammobium, 21
Anchovy industry, South America, 154, 185
Androsace, 42
Animal genetic engineering, 196, 197
Annual meadow grass, 72
Anti-transpirants, to relieve drought stress in plants, 48
Aphids
  adaptable sex life, 31–2
  cabbage root fly as 'unsettler' of, 32
  hover fly as predator of, 34
  sycamore as host for, 95–6
  willow, 112
Apollo Pest Control Ltd, 112
Apple crop failure, 65
Aquifers, 5
Arable farming, 180

Arctic-alpine plants, 56, 228
Arctic charr, 140, 141–2
Argentine osier industry, 189
Arun District, Sussex, wasp plague, 112
Asphodel, 21
Atlantic period, climate of, 11
Aubergines, outdoor cultivation, 22
Audubon, J.J., 165
Aurelia jellyfish, 128
Auriculas, 45

Bacillus thuringiensis (BT), 201
'Balance of nature' in aphid control, 32
Barnacle, Australian, 158
Barnacle geese, 132
Barnet Council, wasp control, 112
Basil, 21
Beans
  field (faba), 185–6
  navy, 186
  soybean, 186
Bee orchid, 81
Beech, dormant buds, 66, 67
Bees, as pollinators, 24
Bellamy, Professor David, 1
Bembridge Ledges, 158
Berkeley, Michael, 125
Berry crops, climate and, 63
Beta vulgaris, 182
Bibury, road verges study, 82–3
Birch
  as host of insects, 95
  seed dispersal, 76
Bird cherry, 57
Birds, coast-dwelling, botulism, 160–1
Birmingham University, plant studies, 70, 183, 184
Black knapweed, 84
Blackthorn, 65
Bloody cranesbill, 134

237

Bluebell, 84
Bogbean, 135
Bog pimpernel, 135
Bosham Creek, sealife in, 126–8
Botanical Society of the British
 Isles, 226
Botrychium lunularia, 136
Botrytis, 23, 25
Bottlebrush, 16
Botulism, in coastal birds, 160–1
Bovine somatropic hormone (BST), 197
Bovine spongiform encephalopathy
 (BSE), 185
Brassica, 138
Brenchley, Patrick, 172
Brent geese, 132
Bronze Age, 11
Brown, V.K., 115
Brussels sprout, search for perfect, 199
BSE (mad cow disease), 185
BST (bovine somatotropic hormone),
 197
BT (*Bacillus thuringiensis*), 201
Burgess Shale rocks, 172
Burnet rose, 134
Bush crickets, 114–15
Butterbur, 84
Butterflies
 Adonis blue, 122
 black hairstreak, 122
 clouded yellow, 120
 comma, 123
 common blue, 121
 dark green fritillary, 123
 effect of climate changes, 118–24
 Glanville fritillary, 121
 grayling, 123
 green hairstreak, 121
 heath fritillary, 227
 holly blue, 123
 large blue, 174
 large copper, 122
 large skipper, 123
 long-tailed blue, 120
 map, 121
 mountain argus, 122
 mountain ringlet, 122
 orange tip, 121
 painted lady, 120
 pearl-bordered fritillary, 119, 123
 purple emperor, 122
 Queen of Spain fritillary, 120
 red admiral, 120
 ringlet, 122, 123
 Scotch, 122
 silver-washed fritillary, 122
 small copper, 121
 small heath, 121
 speckled wood, 119, 123
 tortoiseshell, 119
 white, 120
Butterflies (Professor E.B. Ford), 110

Cabbage family, seawater resistance of,
 138
Cabbage root flies, as 'unsettlers' of
 aphids, 32
Caenorhabditis, 152
Calendula, 21, 34
California, University of, weed
 studies, 72
Callicarpa rubella, 14
Cambridge University, studies
 cat flea, 110
 ladybird, 33
 quinoa, 188
Cambridge Water Company, 4
Canada, Ontario lakes study, 141
Cannabis
 as agricultural crop, 183
 climate changes beneficial to, 11
Carbon dioxide, atmospheric, elevated
 levels of, 152, 206–7
Carboniferous period, vegetation
 of the, 54
Caribbean loggerhead turtle, 156
Carrots, seed germination, 70
Cataclysmic climate changes, effects
 of, 171–3
Caterpillars
 seasonal drought and, 173
 wasp as natural predator, 111
Cats, 1990 flea plague, 110–11
Chalk downland, vegetation of, 81–2
Chemical control of slugs and snails,
 29–30
Cherill, A.J., 115
Chew Valley reservoir, 3
Chichester Harbour, sealife in, 126
Chickweed, 42
Chilean southern beech, 17
Chimonanthus praecox, 19
Chinese mitten crab, 157
Choisya, 18
Church, A.H., 216–17
Cistus, 16

238

INDEX

Cladonia pixidata, 78
Clematis, 17
  cirrhosa, 20
  montana, 46
Clianthus puniceus, 16
Climate changes
  climatic optimum, 11
  cyclical fluctuations in temperature patterns, 11–12
  effects on flora and fauna, 10–13, 64–6
  future course of, 7–9
  gales, 8, 90, 91, 92
  gardening, effect of rising temperatures, 14–26
  late frosts, 16–17
  1980s, record temperatures, 6
  north–south divide in Britain, 9
  plant growth, temperature changes and, 14–26, 64–6
Cloches as frost protection, 46
Clostridium botulinum 160
Club mosses, 55–6
Coastal erosion as result of sea-level changes, effects, 125–36
Coastal flooding, destructive effects, 137–9
Cock's foot grass, 82
Coenagrion damselfly, 116
Collis, John Stewart, 27
Comfrey cultivation, 187
Community Forests, 67
Composting, 47
Conservation
  dilemma surrounding, 174–8
  organizations, 165–6
  soil, 212–13
Consumer psychology, need for change, 200
Convolvulus cneorum, 15–16
Coppices, 89–90
Coreopsis, 34
Corncockle, 217, 218
Cornflower, 217, 218
Corophium, 132
Coulson, John, 117–18
Countryside, future of, 222–36; *see also* Landscape
County Wildlife Trusts, 166
Cowslip, 69, 134
Cow-wheat, 227
Crabs
  Chinese mitten, 157
  hermit, 152–3

Crabbe, George, 223
Crane fly, 118
Cranesbill, 134
Crickets, bush, 114–15
Crimson clover, as green manure, 44
Crouch, Martha, 197
Cupressocyparis leylandii, 40
Curled dock, 214, 215

Daffodil, 45
Dahlia, 16
Daisy family, 82
Damselflies, 116–17
Dandelion, 82
Daphnia, 142–4, 147
Dartmoor, large blue butterfly, 174
Darwin, Charles, 74, 113, 171, 226
Decticus verrucivorus, 115
Deforestation in Bronze and Iron Ages, 11
Dennis, R.L.H., 119–23
Dent, Edith, 218
Developing World, plant genetic engineering and, 197–8
Dionysias, 42
Dormant resting buds, 66–7
Downland vegetation, 81–2
Dragonfly, 117
Drosophila, 151–2
Drought, effects of
  damselfly and dragonfly populations, 117
  downland turf, 81–2
  gardens, 46–8
  mosses and liverwort, 52–3
  water resources, 2–4
  wetland sites, 146–8
Duke University, N. Carolina, honeysuckle studies, 62
Dune gentian, 135
Dunes, *see* Sand Dunes
Dunn, S.T., 59
Durham University
  liverworts and mosses study, 51
  plant transplantation experiment, 228–9
Dust bowls, United States, 212
Dyers' rocket, 214, 215

East Anglia, University of, rising sea levels study, 129
Ecosystem, stability, 74–5

239

Ecuador, changes in winds and ocean
  currents, 153–5
Edwards, Kevin, 183
Elminius modestus, 158
El Nino, sea current, South America,
  154–5, 185
Embleton Golf Links, 134
Energy production, plants for, 188–90
Encarsia wasp, as predator of whitefly,
  37–8
English Nature, 165–6, 228
Ergot, 208–9
Escherichia coli, 178
Esthwaite Water, water flea numbers,
  142–4
Estuaries, threat to bird life in, 133
Estuarine flooding, destructive
  effects, 137–9
Etherington, John, 214, 215
Eucalyptus, 16
Eucryphia, 17
European Community
  agricultural policy, 180, 181, 204
  sorghum studies, 188
Evans, Ceri, 214, 215
Evening primrose, cultivation, 187
Evergreen conifers, dormant resting
  buds, 66–7
Everlasting flowers, 21
Exeter University, moorland fires
  study, 104
Exotic marine flora and fauna, findings
  around British coast, 155–9
Extinctions of organisms as result of
  climatic change, 54, 171–3

Farming industry
  carbon dioxide emissions and, 206–7
  climatic problems, 204–6
  complaints about the weather, 210
  economy, 180
  efficiency of modern methods, 179–90
  European Community and, 180,
    181, 204
  excesses of, 200
  fungal pathogens, 207–9
  future of, climate-wise, 206–9
  recession in, 204–6
  soil erosion and, 210–13
  see also Organic agriculture
Fauna, British, effects of climate
  amelioration, 9–13, 54, 171–3
Fens, butterflies of the, 121–2

Ferns
  adder's tongue, 136
  effect of drought, 53
  moonwort, 136
  sex-life of rusty-back fern, 53
Fertilizers, shoring up soil fertility, 213
Field (faba) bean, cultivation, 185–6
Field vole, 99–100
Fir club moss, 56
Fires, heathland, 103–5, 106
Fish
  exotic types found in British coastal
    waters, 155–8
  freshwater, effect of seawater
    encroachment, 139
Fitter, R.S.R., 56
Flannel bush, 15, 16
Flax, cultivation, 183–4
Fleas, 1990 population explosion,
  110–11
Floating mulches, 46
Flooding, coastal, destructive effects,
  137–9
Flora, Britain
  climate amelioration, effects of, 9–13
  introduction of aliens, 58–62
  north–south divide, 57
Flowering of Britain, The (R. Mabey),
  204
Flowers
  autumn sowing of annuals, 41
  buds, effect of warm spring weather,
    45–6
  hardy perennials, 20–2
Food psychology, 200
Ford, E.B., 110, 113
Forests/Forestry
  conservation policy, 67–8
  results of gale damage, 90–4, 96–7
  sound management essential after
    gale damage, 96–7
Fossil plant deposits, analysis of, 10
Fremontodendron californicum, 15, 16
Freshwater ecosystems, sea flooding
  and, 138–9
Frog
  decline of the, 145–6
  German study, 167–9
From Flint to Shale (M. Berkeley), 125
Frosts
  declining number of, 39, 40
  effects of late, 16–17
  farming industry hit by, 205

INDEX

opening buds and, 67, 68
protection for tender crops, 46
Fruit, exotic soft, successful cultivation of, 23–4
Fruit fly, 151–2
Fruit trees, effect of warm spring weather, 45–6
Fuchsia, 16
Fungal diseases
  farming industry and, 207–9
  garden plants, 42–3, 49
Fungus–plant partnership, 85–8

Gaia, 219
Gaillardia, 34
Gardening
  calendar, rewriting, 39–49
  climatic change, effects of, 14–26
  growing season, lengthening of, 39–40
  micropropagation, 18
  reselection of new varieties, 17–20
  water shortages and, 4
  wildlife gardening, 227
  winter-flowering shrubs, 19–20
  see also Greenhouse gardening
Gazania, 16
Gear, Annabel, 77
Genetic engineering, *see* Animal *and* Plant genetic engineering
Genetic variability, 170–8
Gentian, 228, 229
Germany
  frog study, 167–9
  sunflower cultivation, 184
Giant hogweed, 58
Gladioli, 21, 22
Glasshouse, *see* Greenhouse gardening
Glasswort, 130–31
Glastonbury thorn, 71
Golf links
  coastal erosion and, 134–5
  habitat for wildlife, 135
Gonyaulax, 160
Goodwood Country Park, plant studies at, 81–2
Gould, Jay, 172
Grape hyacinth, 21
Grapes, cultivation, 23–4
Graphis elegans, 78
Grass-of-Parnassus, 135
Grasses, genetic engineering, 194–5
Grasshoppers, 114–15

Grazing animals on chalk grassland, 81
Grazing rye as green manure and weed suppressor, 44–5
Great Gales: of 1987, 8, 90, 91; of 1990, 92
Great Seasons, The (David Bellamy), 1
Great raft spider, decline of, 148
Green-flowered wood sage, 214
Greenfly, 31–2
  hover fly as predator of, 34–5
'Green' genetic plant engineering, 199–202
Greenhouse gardening
  changing climate patterns, effects, 24–5
  design changes, 25
  Encarsia wasp as enemy of pests in, 37–8
  exotic fruit cultivation, 23–4
  insect pests, 36–7
  new crops for, 23–5
  whitefly, 37–8
Green Inheritance (Anthony Huxley), 179
Green manures, 44–5
  soil conservation and, 213
Grey squirrel, 97–8
Grime, Dr Phil, 83–4
Ground-cover crops, 44–5
Groundsel, 42
  six generations produced in a year, 72
Grouse fly, 118
Grouse moors
  accidental fires, 104–5
  management of, 103
Guanay cormorant, 154–5
Guano Ilands, 154
Guardian, The, 150
Guelph, University of, seal plague studies, 163

Haematoma (blood-drop lichen), 78
Halo blight, 186
Hamamelis, 19
Hamsterley, Co. Durham, squirrels, 97
Hardwick, Dr Richard, 186
Hardy perennial plants, 20–22
Hardy, Thomas, 101, 223
Hawthorn, varying response to climate changes, 71
Hayward, Miss I.M., 59
Hazel catkins, effect of weather conditions on, 64–5

Heather moorland
  accidental fires, 103-4, 105
  controlled burning, 103
  flora and fauna of, 102-3
Heatwaves in Britain, effects of, 155-9
Hebes, 16
Hedgehog, as predator of slugs and snails, 29
Hedges
  effect of warmer and wetter winters, 40
  removal of by farmers, 210
Helichrysum, 21
Hellebores, 43
Hemlock, 10
Hemp, cultivation, 183
Henry Doubleday Research Association, pest control research, 29
Henry VIII, King, 183
Herbs, 21
Hermit crab 152-3
Herring gull, botulism, 161
Highland fleabane, 57
Himalayan balsam, 58, 61, 70, 71
Hirnanthian Glaciation, 171-2
Hogan, Paul, 165
Hogweed, 58
Holderness, coastal erosion, 130
Holly, pollination, 63
Holmes, Oliver Wendell, 191
Honeysuckle, 19, 62
Hopkins, Brian, 80-2
Horticulture, *see* Gardening
Hosepipe bans, 3-4, 5
Hosta, 18
Hover fly
  increasing numbers, 34
  pollen-feeding habits, 34-5
  as predator of aphids, 34
Humboldt penguin, decline of, 155
Huntley, Brian, 77
Huxley, Anthony, 179

Ice Ages, 10, 55, 56
Imperial College, wart biter studies, 115
Indian bean tree, 17
Insects
  adaptation to climate changes, 113-24
  greenhouse pests, 31-8
  moorland types, climate change and, 117-18

  see also individual references, e.g. Aphids, Butterflies, Hover fly *etc.*
Institute of Horticultural Research, cabbage root fly studies, 32
Institute of Terrestrial Ecology studies, 129, 169
Intergovernmental Panel on Climate Change, 129, 169
Iron Age, 11
Isle of Wight
  Glanville fritillary butterfly, 121
  Japweed on shores of, 158
  red squirrel, 97

Jackson, Helen, 108-9
Jamaica primrose, 16
Japanese honeysuckle, 62
Japanese knotweed, 58, 61
Japanese marine species in British coastal waters, 157-9
Japweed, 158-9
Jekyll, Gertrude, 61
Jellyfish, 126-7, 128
  Portuguese man-of-war, 156-7

Kaiserlautern, frog populations, 167-9
Kelp, giant, 159
Kettlewell, B.D.H., 113
Knotweed, 58, 61
Kudzu, 62

Lacewing fly, as aphid predator, 35-6
Ladybirds
  increasing numbers, 33
  Phalacrotophora fly as predator of, 33
  as predator of aphids, 32-3
  transformation in appearance, 113-14
Lady's-slipper orchid, 227
Lady's smock, 135
Lakes, Britain
  Arctic charr in, 140, 141-2
  pollution, 141
Lakes, Canada, study of thermoclines, 141
Lancashire, sand lizards, 108-9
Lancaster, Ian, 152-3
Lancaster University, weed studies, 216
Landscape, rural
  postwar changes, 217-18, 220
  preconceptions concerning, 217
Lavender, 21

INDEX

Lavigne, David, 163
Leaves, composting of, 49
Legume cultivation, 184–6
Lepidodendron, extinction of, 54
Lichens, 77–80
  blood-drop, 78
  carbon dioxide and, 79
  changing climate and, 79, 80
  components of, 79–80
  ecological role, 78
  extreme resilience, 78–9
  map lichen, 78
  melting glaciers and, 78
  pixie-cup, 78
  slow growth rate, 78
  source of drought-defeating compound, 48
Lily beetle, 38
Linnaeus, 61, 115
Linseed, cultivation, 183–4
Little Ice Age, 11
Liverpool University, studies
  extinctions, 172
  insects, egg production, 116
Liverworts, 50–4
  effects of drought, 51–3
  sex life of, 51–2
  two varieties, 51–2
Lizard orchid, 135
Lizards, sunshine requirements, 106–9
Lobster claw, 16
Lobsters, octopus population explosion and, 150
Loggerhead turtle, 156
Loch Doon, Artic charr in, 141–2
Loch Ness, Arctic charr in, 140
Long Ashton Research Station, willow studies, 189
Lonicera, 19, 62
Lophocolea cuspidata, 52
Lovelock, J.E., 219
Lowland heaths
  fires, 106
  flora and fauna, 106–9
  grazing animals, 105–6
  Neolithic hunters' forest clearance, 105
  reptiles of, 106–9
Lupins, agricultural varieties, 186
Lyke Wake Walk, Yorkshire, 103

Mabey, R., 204
McLintock, D., 56

Mad cow disease, 185
Mahonica japonica, 19
Mainz University, frog study, 167–9
Maize, cultivation, 184
Majerus, Dr Mike, 33
Manchester Grammar School, 119
Manchester University, isohel study, 108–9
Marine flora and fauna, exotic types found in British coastal waters, 155–9
Marjoram, 21, 215
Markgraf, sugar beet discovery, 182
Marsh helleborine, 135
Marsh orchid, 135
Marsh samphire, 130–1
Mediterranean plants, 21
Melons, cultivation, 25
Metaldehyde, control of slugs and snails, 29–30
Meteorological science, predictive capabilities, 7–8
Micropropagation of plants, 18
Migration of trees, 77
Millponds, disappearance of, 2–3
Molluscs living on mud flats, 132
Moonwort, 136
Moorland insects, effect of climate change, 117–18
Mosses, 50–4
  club, 55
  on downland, 81
  drought conditions, 51–3
  growth on charred heathland, 104
Moth, peppered, 113
Mountain avens, 56
Mountain flowers (M.Walter), 50
Mouse-ear hawkweed, 81, 82
Mud flats
  changing sea levels and, 132
  fauna of, 132
Mulching, 46–8
Mumford, Pauline, 70
Mycorrhizae, role of, 85–8

Napoleon, European sugar industry and, 182
National Rivers Authority, sea defences, East Anglia, 129
National Vegetable Research Station, studies
  navy bean, 186
  tomato, 22
Natterjack toad, 135

Natural history societies, 225–6
Navy bean, 186
Nectarines, cultivation, 24
Neolithic farmers, 214
New England College, Arundel, 80
Newt, decline of, 145
Norfolk
  Broads, coastal flooding, 138–9
  swallowtail butterfly, 124
North Atlantic Drift, role of, 8–9
Northumberland
  butterflies in, 122, 123
  sand dunes of, 134
Nothofagus, 17

Oak
  as host for insects, 95
  late production of first acorn, 72
Octopus plagues, 150–51
Oilseed rape, 181, 182, 183
Ontario lakes, study of, 141
Opencast coal mining, soil erosion and, 211
Ophioglossum vulgatum, 136
Oranges and lemons, cultivation, 24
Orchids, wild, 81, 135, 227
Oregon State University, mycorrhizae studies, 85–7
Organic agriculture
  insurance against soil erosion, 213
  and plant genetic engineering, 199–203
Origanum vulgare, 215
Origin of Species, The (C. Darwin), 74, 113
Osier production, 188–90
Otters, 227
Owls, as predators of field voles, 100
Oxford Polytechnic, 119
Oxford ragwort, 61
Oyster plant, 57
Oysters, Japanese, 158

Pacific marine species in British coastal waters, 157–9
Parry, Dr Martin, 184
Peaches, cultivation, 24
Peat bogs, 10–13, 47
  sub-fossil pollen deposits, 10–11, 57, 183
Pelargonium, 16
Penguin, Humboldt, 155
Penstemon, 16

Penwith Sixth Form College, hermit crab study, 152–3
Peppered moth, 113
Peripatetic plant species, concept of, 76
Permian period, vegetation in, 54
Perry, Dr Dave, 85–7
Peru, changes in winds and ocean currents, 153–5
Pesticides, use of, 29–30
Pests in garden and greenhouse
  new types of, 38
  temperature rises, effect of, 31–8
  warmer, wetter winters, effect on insects, 43
  see individual references, e.g. aphids, slugs, snails *etc.*
Pests, marine, flora which have become a problem, 158–9
'Peter Grimes' (G. Crabbe), 223
Phacelia
  as green manure, 44
  pollen of, 35
Phalacroptophora fly, as predator of ladybirds, 33
Physalia physalia, 156–7
Phytophthora infestans, 207–8
Phytotron, use of in plant breeding, 192
Pilchard, 157
Pineapple weed, 61
Plankton, coastal waters, 126–8
Plant genetic engineering, 191–203
  defect correction, 194–5
  Developing World and, 197–8
  experimental approaches, 192–3
  'magic bullet technique', 195
  marketing the technology, 195–9
  sensitivity of plants to daylength, 193
  switch to organic methods, 199–203
  trial sites overseas, 192–3
Plants, Britain
  aliens, introduction of, 58–62
  annuals, behaviour as winter approaches, 68
  cell division and flowering, 83–4
  changes in status as result of climate change, 169–70
  evolutionary changes as result of climate change, 72
  introduced species, 57–62
  length of life cycle, 72
  life through the ages, 10–13
  perennials, behaviour as winter approaches, 68

INDEX

pond margin species, 145, 146–7
significant association with soil organisms, 85–8
transplantation experiment at Upper Teesdale, 22–9
water loss from leaves, control of, 48
see also Trees *and* Vegetation
Plum crops, late frosts and, 16
Pocket Guide to Wild Flowers, The (McLintock and Fitter), 56
Pollen
plants providing plentiful source of, 34–5
sub-fossil deposits, 10–11, 57, 76, 183
Pollen-feeding insects, 34–5
Pollination
bees' role, 24
blackthorn, 65
crops, 24
effect of poor conditions on seed set, 64
hazel catkins, 64–5
Polyanthus, 45
Ponds
effect of climate change on, 145–7
pollution, 147
Porter, Dr John, 189
Portsmouth Polytechnic, Japweed studies, 158
Portuguese man-of-war, 156–7
Potato blight, 207–8
Potatoes, cultivation, 193
Prairie-farming, 210, 211
Prawns, Japanese, 158
Prehistoric extinctions, 171–3
Primrose, Teesdale, 228, 229
Primulas, 43, 45
Pruning
revision of programme for, 40
roots, 40–41
roses, 40
shoots, 40, 41
Prunus serrula, 15
Pueraria lobata, 62
Puffins, effect of sand eel decline, 151
Purple loosestrife, 135
Pyramidal orchids, 134

Quinoa, cultivation, 187–8

Rackham, Oliver, 89
Ragworms, 132
Rainfall, decrease in, 3–6

Recycling, mulching material as result of, 47–8
Red clover, 44
Red mullet, 156
Redshank, 132
Red spider mite, 36–7
Red squirrel, 97–8
Reptiles, lowland heaths as habitat of, 106–9
Reseda luteola, 214
Research into climate change, funding, 224–5
Reservoirs, pollution, 147
Return of the Native (T. Hardy), 101
Rhizocarpon geographicum, 78
Rhodanthe, 21
Rhododendron
dauricum, 19
ponticum, 60, 61
Ribwort plantain, 82
Rivers
Derwent, 5
drying up of, 3, 5
Frome, 5
Hamble, 5
Lavant, 3
Little Stour, 5
Meon, 5
Thames, ice fairs on, 11–12
Wharfe, 5
Road verges, vegetation on, 82–3
Rosedale Head, Yorkshire, 104
Rosemary, 21
Roses, effect of warmer winters, 40
Rothamsted Experimental Station, molluscs study, 30
Rottingdean, soil erosion incident, 211
Rough hawkbit, 82
Roundworm, 152
Rove beetle, 118
Royal Society for Nature Conservation, 166, 226
Royal Society for the Protection of Birds, 226
Royal Troon, 134
Rumex crispus, 214
Rusty-back fern, 53

St Andrews, University of, 183
Salt marshes
changing sea levels and, 125–33
fauna of, 132
life on, 125–8

245

threatened flora, 130–2
Saltwort, 131
Salvia, 16
Sand dunes
  dune slack, habitat for rare plants, 135
  rising sea levels and, 134–6
  threatened flora of, 134–6
Sand eels, deline of, 151
Sand lizard, 107–9
Sargassum muticum, 158–9
Sasek, T., 62
Saxifrage, 42
Science and the future of the countryside, 222–3
Scotland, cannabis and hemp cultivation, 183
Scots pine, migration of, 77
Scurvy grass, 130, 131
Sea aster, 131
Sea currents, change in, 153–9
Sea horse, short-snouted, 157
Sea kale, 133
Sea lavender, 130, 131
  cultivated version, 187
Sea level changes, the seashore and, 125–36
Sea pink, 229
Sea purslane, 131
Sea rocket, 134
Sea spider, 157
Sea spurrey, 131
Sea squirt, 157
Seal plague, effect of, 161–4
Sedgemoor, osier beds, 189
Seeds
  biochemical mechanism in, 69
  carrot, germination, 70
  methods of distribution, 59–61, 76–7
  sycamore, 95
  types needing prolonged cold before germination, 43, 69–71
Sex life of
  mosses and liverworts, 50–4
  rusty-back fern, 53
Sheep's fescue, 81, 82
Sheffield University, plant studies, 83
Shingle beaches, effect of sea-level changes, 133
Shredders, recycling pruning by means of, 47
Shreeve, T.G., 119–23
Shrimps, 132

Shrubs, early-flowering, frost and, 65
  winter-flowering, 19–20
Silene acaulis, 42
Silt as source of water pollution, 211
Simpson, Gordon, 97
Sites of Special Scientific Interest, 129, 135, 158, 224, 230
Sitka spruce, dormant buds, 66, 67
Slender speedwell, 59, 61
Slow worms, 106–7
Slugs
  Abax, as predator of, 29
  chemical control methods, 29–30
  feeding patterns, 28
  as garden pest, 27–31
  hedgehog as natural enemy of, 29
  migration to deeper soil levels, 28
  survival in dry conditions, 28
Small-leaved lime, 57–8
Snails
  aestivation, 28
  chemical control methods, 29–30
  feeding patterns, 28
  as garden pest, 28–31
  pampas grass as a hideout for, 28
  self-sealing of shells, 28
  survival in dry conditions, 28
Snowdonia National Park, *rhododendron ponticum* in, 60
Snow moulds, 49
Soil erosion, 210–13
  methods to combat, 212–13
  organic farming and, 213
Soil fertility, building up, 43–5, 213
Soil organisms and plants, significant association between, 85–8
Somerset, Glanville fritillary butterfly in, 121
Somerset Levels, rising sea levels and, 130
Sorghum, cultivation, 188
South America
  anchovy industry, decline, 154, 185
  changes in winds and ocean currents, 153–5
Southampton University, studies lizards, 107
Phacelia, 35
Southport, sand lizard colonies, 108–9
Soybean, cultivation, 186
Spartina grass, 125, 126
Speedwell, 59
Spellerberg, Dr Ian, 107–8

INDEX

Spencely, T., 103
Spider mites, 36–7
Spiders
  decline of Great raft spider, 148
  sea spider, 157
Spirea x arguta, 17
Spring gentian, 56
Springs, earlier and shorter seasons, 45–6
Squirrels, 97–8
Stachys lanata, 18
Stag's horn club moss, 56
Statice, 21, 187
Steep Holm, botulism in herring gulls, 161
Stereocaulon, 78
Strain, Boyd, 62
Succulent glasswort, 130–1
Suckling clover, 81, 82
Suffolk Trust for Nature Conservation, 148
Sugar beet, 181, 182
Sunfish, 156, 157
Sunflower, cultivation, 184
Sweetcorn, 22
Sweet peppers, outdoor cultivation, 22
Sycamore
  as host for insects, 95
  seed dispersal, 76
  spread of the, 94–6
Symondson, W., 29

Tar-spot fungus, sycamore as host for, 95–6
Teesdale, plant transplantation experiment, 228–9
Terns, 133
Tettigonia viridissima, 114
Teucrium scorodonia, 214
Thermoclines of deep-water lakes, 140, 141
Thompson, D., 116
Thresher sharks, 155
Thrift, 131
Thyme, 16, 21
Tidal surges, destructive effects, 138–9
Toad
  decline of the, 145
  natterjack, 135
Tobacco plant, experiments with, 195
Tomato, outdoor cultivation, 22
Towyn, flooding of, 137, 138
Transplantation of plant communities, experiment, 228–9
Transport infrastructure, effect on wildlife, 167
Trees
  conservation policy, 68
  dormant resting buds, 66
  early-flowering, weather and, 64
  gale damage, 90–4, 96–7
  leaf shedding, 66
  low-temperature requirements for bud burst, 66–67
  migration of, 77
  mild winters, effect of, 66–7
  planting of, as fashionable trend, 67
  seed dispersal, 76–7
  weather conditions, effect on growth, 64
Trefoil, 44
Trees and Woodland in the British Landscape (O. Rackham), 89
Trigger fish, 156, 157
Trilobites, extinction, 172
Tube worm, 157
Tulip, 21
Tulip-tree, 10
Tunny, 157
Twite, 132
Tyne and Wear, wildlife corridors, 231

United States, dust bowls, 212
Uplands, flora and fauna of, 101–3
Upper Teesdale, transplantation experiments, 228–9
Usnic acid, 48

Vegetables
  effect of climate changes on growing of, 22–3
  extended growing seasons, 23
  low-temperature tolerance, 41
  winter sowing, 41
Vegetation
  classification of different types of, 75–6
  climate-induced changes in, 84

247

downland, 81–2
heathland, 102–3
upland, 101–3
see also Flowers, Plants, Trees, *etc.*
Victoria amazonica, 155
Vines, introduced varieties, now a weed nuisance, 62
Violets, 43
Teesdale, 228
Viviparous lizard, 108
Vole plagues, 99–100

Wales, University of, studies
Abax, 29
weeds, 214
Walker, B., 97
Walnut trees, frost and, 17
Walters, M., 50
Wart biter, 115–16
Wasps
high summer temperatures beneficial to, 112
nest sites, 112
1990 population explosion, 111–12
as predator of aphids, 111, 112
Water
increasing domestic demand for, 5
levels, rapid fall of, 5
meters, increasing use of, 5
pollution, 147, 211
resources, over exploitation, 5
shortages, 3–6
Water flea, 142–4, 147
Water plantain, 135
Weather, *see* Climate changes
Weeds
climate changes and, 42, 72, 213–17
heyday of, 216
introduced types, 60–2
loss of arable varieties, 217
moisture stress and, 214–15
seeds, germination, 214–16
Wetland habitats
climate change and, 145–9
management needed to preserve, 148–9
White, Gilbert, 14, 25, 226
Whitefly
Encarsia wasp as enemy of, 37–8
as greenhouse pest, 37–8
rising temperature beneficial to, 37
Russian research into control of, 38
sticky yellow traps for, 38
Whitethroat, 89
Whittington, Graeme, 183
Wild Flower Magazine, 218
Wild Flower Society, 218
Wildlife
conservation, 227–34
gardening, 227
possible benefits from climatic change, 221
provision of corridors for, 230–4
Wildlife and Countryside Act 1981, 59, 115
Willis, Professor Arthur, 82–3
Willow aphids, 112
Willow, coppiced, cultivation, 189–90
Wind
effect on surface water temperature, 140–1
forest damage from, 90–4, 96–7
Wingnut, 10
Winter field beans, 44
Wintergreen, 135
Winter-sweet, 19
Winter tares, 44
Winters, milder and wetter seasons, 39–40
Witch hazel, 19
Witton Moor, North Yorkshire, fire, 103
Wonderful Life (Jay Gould), 172
Wood sage, 214, 215
Wratten, Dr Steve, 35

Xeranthemum, 21

Yellow-horned poppy, 133